ORGANIC CHEMISTRY:

MAXIMUM SUCCESS

MINIMUM EFFORT

PART I

Professor Jakob Fredlos

Table of Contents

Organic Chemistry: Maximum Success, Minimum Effort

Preamble

If you are reading this book, I hope you are taking Organic Chemistry, or have some reason to want to know about Organic Chemistry. (If not, why are you reading this book? Get a life!) Odds are, you are a pre-medical college student or in some similar program, and you are wondering "WHY?" Why do you have to take this class at all? You aren't planning to be a chemist, much less an organic chemist, so what use does this information have for you? There are two answers to your question. First – stop whining! it doesn't matter WHY. The medical school admissions committees have decided that Organic Chemistry is a good test of the kind of intelligence you will need for medical school, so they will continue to use it in their own quasi-Darwinian way: survival of the fittest – and you want to be one of the fittest! Second, the information is a *little* bit useful – if you are successful in your career quest, you will likely prescribe drugs, and most of them are organic chemicals; and you will apply some biochemistry, and that is largely based on organic chemistry. That makes structure, nomenclature and reactivity a little useful. Still, understanding spin-spin coupling in NMR, anti-aromaticity or the molecular orbitals of conjugated polyenes is a little extreme – it is doubtful that those will ever come in handy later in life, unless you actually become a chemist!

So what you want is a guide that points you to the *important* parts of Organic Chemistry, shows you what to study, what to memorize, what to know. Gives you tips about the types of problems you might see on the exams, and helps you get the grade you need with the least amount of effort.

You have come to the right place. As an experienced Professor of Organic Chemistry, I know what is important, what to study, and what to memorize – and the tricks that the Organic Professors love to play on unsuspecting students!

Cognitive scientists have shown that we can remember up to 7 things easily. Anything over 10 is hard. That is why we break up long numbers we need to remember- 3 digit area code and 7 digit phone number; or your social security number as xxx-yy-zzzz. To aid in remembering, for each topic, we will have at most 7 points. If we need more, we will break the topic up into separate sections.

So let's get started.

Professor Jakob Fredlos

CARBON: The Beginning

Your professor will delight in telling you that Organic Chemistry is the chemistry of Carbon. They may provide a long-winded explanation of how the names Organic (= from living organisms) and Inorganic (= not from living organisms) came about, how Organic came to mean Carbon, and mention how "Biochemistry" (= chemistry of life processes) split off from Organic. There may be a bad joke about Organic chemistry and Organic food. You may hear about a chemist named Wohlers who first made Organic (urea) from Inorganic (air – carbon dioxide and nitrogen) with a snickering aside about the source of urea (urine), and therefore disproved the theory of the *"elan vital"*. Smile and nod. The history lesson probably won't be on the test.

But here is what is important: Of something like 10 million known chemical compounds, 9 million or more have carbon in them and can be considered organic. Most of the compounds and processes vital to living organisms are based on carbon. So that is one reason why we study it, and why carbon gets to have a branch of chemistry all its own. Still, it is unlikely any of this is on the test.

I.   Structure and Bonding

1.   Atomic Structure, Atoms, and Elements

Almost every Organic Chemistry textbook starts out with Atomic and Molecular Structure. A lot of this was taught in General Chemistry in a traditional program. But that doesn't mean you learned it then OR that you remember any of it now! So here are the things you need to know.

1. *Atoms.* All matter (everything that isn't energy!) is made of **atoms**. Just accept this. Make it a fundamental assumption of your Organic Chemistry reality. Atoms are REALLY REALLY REALLY small – a teaspoon full of sugar has 4×10^{23} atoms, nearly one mole. If you counted all of the grains of sand on all the beaches on the entire Earth, the number of grains is less than 10^{19}. So in your teaspoon of sugar, the number of atoms is more than 10,000 times greater than all the grains of sand on Earth – now THAT is small!

2. *Elements.* Atoms come in different types called **elements**. Think of this like cars or shoes – they come in different "makes" or brands – Audi, BMW, Chevrolet; Nike, Puma, Adidas etc. Each element is defined by an ATOMIC NUMBER (the number of positively charged particles called protons in the nucleus) and given a name and symbol.

3. *Periodic Table.* You can understand a lot about the elements from the **Periodic Table**. We don't need to get into the depths of it, but let's take a quick look at some things.

Here is the way the Table usually looks:

PERIODIC TABLE OF THE ELEMENTS

IA 1	IIA 2	IIIB 3	IVB 4	VB 5	VIB 6	VIIB 7	VIIIB 8	VIIIB 9	VIIIB 10	IB 11	IIB 12	IIIA 13	IVA 14	V 15A	VIA 16	VIIA 17	VIIIA 18
H																	He
Li	Be											B	**C**	**N**	**O**	**F**	Ne
Na	Mg											Al	Si	P	**S**	**Cl**	Ar
K	Ca	Sc	Ti	V	Cr	Mn	Fe	Co	Ni	Cu	Zn	Ga	Ge	As	Se	**Br**	Kr
Rb	Sr	Y	Zr	Nb	Mo	Tc	Ru	Rh	Pd	Ag	Cd	In	Sn	Sb	Te	**I**	Xe
Cs	B	La	Hf	Ta	W	Re	Os	In	Pt	Au	Hg	Tl	Pb	Bi	Po	At	Rn
Fr	Ra	Ac	Rf	Db	Sg	Bh	Hs	Mt									

A little scary, right? Here is the good part: THE SHADED AREAS DO NOT MATTER <u>AT ALL</u> FOR ORGANIC CHEMISTRY! I don't think I have ever seen any of these elements mentioned in a sophomore level book. AND: ONLY THE ELEMENTS IN BOLD ARE REALLY IMPORTANT: H, C, N, O, F, S, Cl, Br, I. So really we are talking about learning about NINE types of atoms. Can you name nine brands of cars or shoes? Nine sports teams? Nine celebrities? I think you can handle nine. Anyway, the Table is usually provided.

The periodic table should be your friend: let us look up close at the box for Carbon.

IV-A 14	These are the labels for this column or "Group" The old name was Group IV-A. (Roman numeral IV) Now it is Group 14. The atoms in this group have 4 valence shell electrons, and most of them make 4 bonds.

Carbon 6 **C** 12.01	The name of the element, the atomic number (6), the symbol (C) and the atomic mass (12.01)

4. *The Important Nine.* The names of the elements are mostly random – you can read about it elsewhere if you really want to know. The symbols are one or two letter acronyms. (A few don't make sense – Na for Sodium? – if you speak Latin you call it Natrium.) You should know the names and symbols of the 9 important elements – it would be good to know the other 17 that are sometimes used in Organic Chemistry.

Important Nine

Other Seventeen

Element	Symbol	Atomic Number	Element	Symbol	Element	Symbol
Hydrogen	H	1	Lithium	Li	Chromium	Cr
Carbon	C	6	Boron	B	Manganese	Mn
Nitrogen	N	7	Sodium	Na	Iron	Fe
Oxygen	O	8	Magnesium	Mg	Nickel	Ni
Fluorine	F	9	Aluminum	Al	Copper	Cu
Chlorine	Cl	17	Silicon	Si	Zinc	Zn
Bromine	Br	35	Phosphorus	P	Osmium	Os
Iodine	I	53	Potassium	K	Mercury	Hg
Sulfur	S	16	Titanium	Ti		

5. *Sub-Atomic Particles.* Atoms are made of smaller particles: protons, neutrons, electrons. In atomic mass units, protons and neutrons have a mass of 1, electrons have a mass 1800 times smaller, so we round down to zero. Protons have a charge of +1, neutrons 0, and electrons -1. The protons and neutrons are packed close together in the center of the atom (the NUCLEUS) while the electrons fly around outside. The ATOMIC NUMBER is the number of protons in the nucleus. So all carbon atoms have 6 protons. An equal number of electrons are outside to make the atom neutral (+6 for protons, -6 for electrons = 0). There are also neutrons in the nucleus for all elements except hydrogen. To keep the nucleus stable (not radioactive), there must be between 1 and 1 ½ neutrons for each proton. The ATOMIC MASS is the combined number of protons and neutrons, since each has a

mass of 1. Individual atoms have masses that are very close to round numbers; the masses listed on the periodic table are average masses. Since atoms of a given element can vary in the number of neutrons (these are *isotopes* of the element), this reflects the typical composition of a sample of those atoms. A typical sample of carbon, for example, is about 99% mass 12, about 1% mass 13, and a tiny trace of the radioactive mass 14. So the average mass of a carbon atom is 12.01 rather than 12.00.

6. *Valence Electrons.* While the type of atom is controlled by the protons in the nucleus, the chemical nature of atoms is controlled by the electrons. The electrons organize themselves around the nucleus in layers, which correspond both to approximate position and to energy. Closer in is lower energy, further out is higher energy. For a given atom, the highest energy layer determines the kind of chemistry it has. We call these the VALENCE ELECTRONS. The periodic table is laid out in such a way so elements with the same number of valence electrons are in the same column! The first column (Group IA in the old system, Group 1 in the new) is all the elements with 1 valence electron. These atoms tend to lose that electron and become ions with a +1 charge, forming salts (Na as Na^+ in NaCl). So they all have similar chemistry based on a similar number of valence electrons. The same is true of the other columns.

7. *The Octet Rule.* As we will see next, most of the atoms like to interact with each other. The only ones that don't are the inert gases on the far right of the periodic table. About a century ago, Professor G.N.Lewis postulated that the best situation for an atom is to be like an inert gas atom. That means having eight (8) electrons in the valence shell – an octet. Since the atoms in a group (column) have the same number of valence electrons, they tend to do the same things to get to the octet: losing one or two electrons, gaining one or two electrons, or sharing electrons.

A lot of what we have covered in this chapter is considered background – you probably won't get test questions about the octet rule or elements or sub-atomic particles. But you will be expected to know it and to build on that knowledge to understand other things. For example, in stable compounds carbon almost always forms 4 bonds (reactive carbon species may have 2 or 3 bonds). This is a result of the four electrons in the valence shell.

Quick Review.

1. ATOMS: Everything is made of atoms.
2. ELEMENTS: Atoms come in different kinds – the elements. There are over 100, only 9 are important.
3. PERIODIC TABLE: The kinds of atoms can be organized using the periodic table
4. NAMES: Each kind of atom has a name and a symbol.
5. PARTICLES: Atoms are made of protons, neutrons, and electrons. The number of protons in the nucleus determines which atom it is, the Atomic Number
6. VALENCE ELECTRONS: The number of electrons in the outer shell of electrons determines chemistry.
7. OCTETS: The preferred state is to have a full valence shell with 8 electrons – atoms will gain or lose or share electrons to attain this full shell.

2. Bonding and Molecular Structure

1. *Chemical Bonds.* Atoms combine with each other to make compounds. We call the links between atoms CHEMICAL BONDS. The bonds are of two types: Ionic Bonds and Covalent Bonds. In **ionic bonds**, electrons are transferred between atoms. One atom gives up an electron or two (rarely three or more), while another atom takes the electrons. When the atoms gain or lose electrons, they become charged particles called IONS. The atom that loses electrons becomes positively charged and is called a CATION. The atom that gains electrons becomes negatively charged and is called an ANION. In the ionic bond, the positive and negative ions are attracted by electrostatic force. So Sodium (Na) gives an electron to Chlorine (Cl) and they become Na$^+$ and Cl$^-$, or table salt NaCl. You should remember this, but ionic bonds are not very important in organic chemistry. Here is ionic bonding as shown with Lewis Dot Structures.

$$\text{Na} \cdot + :\ddot{\text{C}}\text{l} \cdot \longrightarrow \text{Na}^+ :\ddot{\text{C}}\text{l}:^-$$

Sodium gives an electron to chlorine to become sodium chloride with an ionic bond.

2. *Covalent Bonds.* Covalent Bonds are the important bonds for Organic Chemistry. In a **covalent bond**, atoms share electrons. When two atoms share a pair of electrons, each atom gets to count both shared electrons as part of its valence shell. So in the simplest case, two hydrogen atoms come together. Each brings one electron, and the two electrons are shared. So each hydrogen has a full first shell (2 electrons)! Because the electrons are spread out over a larger space (2 Hydrogen atoms instead of 1), they are at lower energy – the reduced energy holds the H atoms together – a covalent bond. Carbon can join together with 4 hydrogens. The carbon has 4 valence electrons, the hydrogens each bring 1 for a total of 8. So the carbon has 8 electrons – a full shell; and each hydrogen has the 2 that it needs for its full shell. Carbon commonly forms covalent bonds with other carbon atoms, with hydrogen, nitrogen, oxygen, and fluorine, with silicon, phosphorus, sulfur and chlorine, with bromine and iodine.

$$\text{H} \cdot + \text{H} \cdot \longrightarrow \text{H} : \text{H}$$

Hydrogen atoms share electrons to form a covalent bond.

3. *Polar Bonds.* Although electrons are shared in covalent bonds, they are not necessarily shared equally. If the sharing is relatively equal, the bond has no charge across it, and we call it a NONPOLAR bond. If the sharing is unequal, then one end will be negative (more electrons) and one end will be positive (less electrons), we call the two ends poles, and the bond a POLAR bond. There is a way of assigning atoms a number that measures their ability to attract electrons – called "Electronegativity". The original electronegativity scale runs from 0 to 4, with the most electronegative atom (Fluorine) given a 4. In this electronegativity scheme, bonds where the electronegativity difference is less than 0.5 are considered NONPOLAR; electronegativity differences greater than or equal to 0.5 are POLAR. Here are the electronegativities of the most important atoms – C 2.5, H 2.1, N 3.0, O 3.5, F, 4.0, Cl 3.5, Br 3.0, I 3.0. Note that this makes C-C and C-H bonds NONPOLAR, while *all other bonds to C and H are polar.* You can learn all the electronegativities or just memorize that C-C and C-H are NONPOLAR while all the other common bonds are POLAR.

4. *Shapes of Molecules.* We predict the shapes of molecules using a technique called Valence Shell Electron Pair Repulsion theory – VSEPR for short, usually pronounced "VESPER" (since VSEPR is unpronounceable). The concept of VSEPR is simple. Imagine the electron pairs in the bonding level (the Valence Shell) are anchored to the atom, but are free otherwise to move around. Since the electrons repel, they try to move as far from all other electron pairs as possible. So if there are 2 pairs of electrons, they move to opposite sides of the atom and make a linear arrangement. If each is connected to another atom as a covalent bond, then the two bonds are LINEAR, with a bond angle of 180°. If there are 3 pairs of electrons, the farthest they can get from each other is to stay in the same plane and point at the vertices of an equilateral triangle, with 120° angles between the bonds. We call this TRIGONAL PLANAR. For 4 pairs of electrons, we move into 3 dimensions, with the bonds pointing at the corners of a tetrahedron, with 109.5° angles, and we call the shape TETRAHEDRAL. These are easier to see than to describe, so here are pictures:

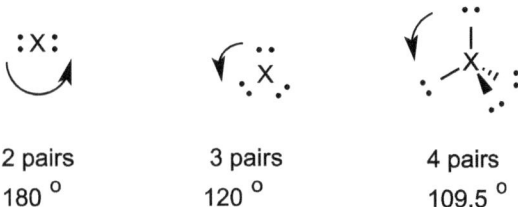

| 2 pairs | 3 pairs | 4 pairs |
| 180° | 120° | 109.5° |

Note: in our picture of the tetrahedron we have to represent 3 dimensions. So the line that is a little wedge extends forward, out of the page; the line that is dashed extends backward, behind the page. Remember this – it will be used quite a bit!

5. *Polar and Nonpolar molecules.* Bonds can be POLAR or NONPOLAR; so can molecules. The molecular dipole is made by adding up the bond dipoles as vectors. There are ways to measure the molecular dipole. But it can be predicted from two easy questions: First, does the molecule have polar bonds? If NO – the molecule is NONPOLAR. (If all bond dipoles you add are zero, you get a zero molecular dipole.) If YES, ask the second question: are the bond dipoles arranged symmetrically so as to cancel? If YES, NONPOLAR. If NO, then the molecule is POLAR. Here are a few examples. The dipoles of polar bonds are shown as arrows pointing to the negative end. The most difficult one to see is the CCl_4 – the 4 polar bonds are arranged symmetrically and cancel. Be able to identify polar bonds and polar molecules – your professor will probably ask you to do this!

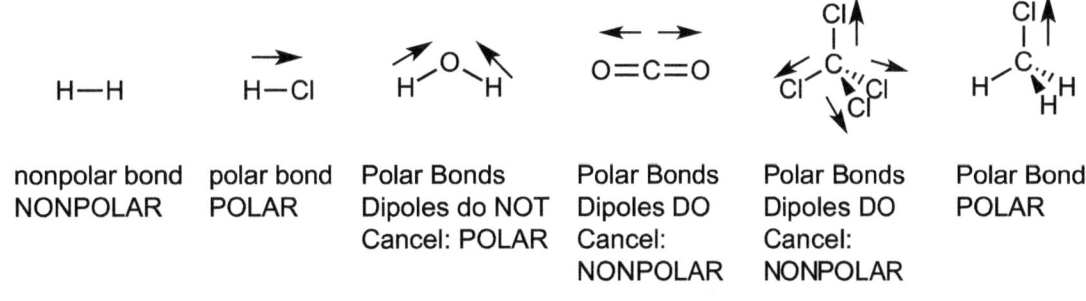

| nonpolar bond
NONPOLAR | polar bond
POLAR | Polar Bonds
Dipoles do NOT
Cancel: POLAR | Polar Bonds
Dipoles DO
Cancel:
NONPOLAR | Polar Bonds
Dipoles DO
Cancel:
NONPOLAR | Polar Bond
POLAR |

6. *Dot Structures.* Earlier we told how the chemist G.N.Lewis formulated the idea of stable octets. He also devised a simple way of keeping track of electrons on atoms as they form ionic and covalent bonds. He used dots to represent the outer shell (valence shell) electrons. So sodium in group 1 has 1 electron in its valence shell so we

write Na$\cdot$ for a sodium atom. If it loses its electron and becomes an ion, it is Na$^+$. Chlorine has seven electrons in its valence shell, so we write $\overset{\cdot\cdot}{\underset{\cdot\cdot}{Cl}}\cdot$ for a chlorine atom. When it takes on an electron (say from sodium) it becomes a chloride anion $\overset{\cdot\cdot}{\underset{\cdot\cdot}{Cl}}{:}^-$. We can represent sharing electrons also. Here is a hydrogen molecule with 2 electrons shared between the hydrogen atoms H:H ! These are called Lewis Dot Structures. As a first approximation, they are a great way to represent ions, atoms, and molecules – but they are a representation, not reality. It is a handy way to see if atoms have the preferred 8 electron octets! Here are a few more Lewis Dot Structures. Occasionally your instructor may request you draw a Lewis Dot Structure – it should look like these!

$$H\ \overset{\textstyle H}{\underset{\textstyle H}{:\overset{\cdot\cdot}{\underset{\cdot\cdot}{C}}:}}\ H \qquad\qquad :\overset{\cdot\cdot}{\underset{\textstyle H}{O}}{:}H \qquad\qquad H\overset{\textstyle H\ \ H}{\underset{\textstyle H\ \ H}{:\overset{\cdot\cdot}{\underset{\cdot\cdot}{C}}:\overset{\cdot\cdot}{\underset{\cdot\cdot}{C}}:}}H$$

In organic chemistry, there are a lot of covalent bonds. Drawing all those dots is tedious. So we will sometimes replace a pair of dots by a line. Then it is a line structure.

$$H{-}\overset{\textstyle H}{\underset{\textstyle H}{C}}{-}H \qquad\qquad :\overset{\cdot\cdot}{\underset{\textstyle H}{O}}{-}H \qquad\qquad H{-}\overset{\textstyle H\ \ H}{\underset{\textstyle H\ \ H}{C{-}C}}{-}H$$

But drawing all the lines is still tedious, so we will NOT draw out bonds to hydrogen, but condense each carbon-hydrogen group down to a molecular formula type representation: CH_3, CH_2, CH, etc. This greatly reduces the number of lines we have to draw – these are called condensed structures. These condensed structures are the most common way to see organic molecules. In the days before molecular drawing programs and personal computers, it allowed people to write structures with a typewriter: $CH_3{-}CH_2{-}CH_2{-}CH_3$ for example.

$$H_3C{-\!-}CH_2{-}O{-\!-}CH_3 \qquad\qquad H_3C{-}CH_3$$

For complex structures, even the condensed structures can be a bit tedious, so there is one more level of simplification. Here we omit all C and H atoms. A line segment represents a C-C bond, so we know the end of any line segment is a C. It always has 4 bonds, so any unused bonds go to H. Other atoms (N, O, F, etc) are shown by their symbols. This is called a skeletal structure. We will use these sometimes, as will your teacher – most often when the organic structure has a ring. You need to be familiar with all of these different ways to represent molecular structures!

7. *Resonance.* If we want to understand molecules at a deeper level, we have to go beyond the Lewis Dot Structure and the other drawing motifs derived from it (line, condensed, skeletal are all equivalent to a Lewis Dot Structure). Why? Because the Lewis Dot Structures fail for certain types of molecules and are always an oversimplification. The basic reason for this is that the Lewis structures give the impression that the electrons are localized – they exist in certain places. But really, the electrons are so small and fast moving that this is not the way they behave – they are able to "delocalize" or spread themselves around. Consider the carbonate anion, CO_3^{2-}. A Lewis Dot structure has a central C and 3 O atoms around it. One has a DOUBLE BOND (two pairs of electrons

shared from C to O – more on this soon) and the O is neutral, while the other two have single bonds from C to O and the O's are charged (-1 each). So this structure predicts one short C=O bond and two longer C-O bonds, and a large dipole moment from the two charges. BUT THIS IS WRONG! Measurement of the carbonate ion shows 3 equal length bonds and NO dipole moment! Why? Because the electrons are not constrained by our Lewis Dot device – they flow like water to the best arrangement they can get. And that has equal distribution of electrons and charge on all 3 oxygens. One way to represent this is through a method called RESONANCE. In RESONANCE, we draw all possible and reasonable electron arrangements using Lewis Dot Structures. For carbonate, that means 3 arrangements where a different O has the double bond. Then, we say that these are RESONANCE STRUCTURES (another name is CANONICAL STRUCTURES) and that the REAL structure is a HYBRID (think average) of all of them. So, each carbon oxygen bond is like 2/3 single and 1/3 double; and each O has 2/3 of a negative charge: now this is consistent with reality: equal bonds, equal charge. Whenever we can draw more than one reasonable structure, RESONANCE will be in effect. We show resonance with double headed arrows. (But be CAREFUL: double headed arrows ←→ mean RESONANCE, 2 arrows in opposite directions ⇌ mean EQUILIBRIUM. In equilibrium, two compounds continuously change back and forth from one to the other. In resonance, the structures we draw are NOT REAL – they are the best we can do with Lewis Dot structures. The real structure is the hybrid.)

When all resonance structures are equivalent – like in carbonate – they all make the same contribution to the real structure – equal amounts of charge on all 3 oxygens. But this will not always be the case. The structures may not be equivalent – and then they won't make equal contributions. The "better" structure makes a larger contribution – the real structure is more like the "better" structure. In these situations, use your common sense: more bonds are better than less bonds, better to have octets than not, better to have (-) charge on O vs. N vs. C (electronegativity), better to have (+) charge on C vs. N vs. O, and so on. You should DEFINITELY know how to draw resonance structures!

Quick Review

1. BONDS: Atoms connect through bonds to make compounds. When electrons are transferred completely to make charged particles or ions, they are called IONIC BONDS.

2. COVALENT BONDS: When electrons are shared between atoms, these are COVALENT BONDS. Covalent bonds are more important for organic chemistry.

3. POLAR BONDS: Electrons may not be shared equally. If equal – NONPOLAR BONDS; if unequal – POLAR BONDS.

4. VSEPR: We can predict the shapes of molecules by VSEPR.

5. POLAR MOLECULES: Molecules can be NONPOLAR or POLAR. If all bonds are NONPOLAR – molecule is NONPOLAR, no matter the shape. If there are some POLAR bonds we must evaluate the shape – if bond dipoles cancel, the molecule can be NONPOLAR. If they do not cancel, it is POLAR.

6. DOT STRUCTURES: Lewis Dot structures, line structures, condensed structures, and skeletal structures are handy representations of molecules showing the electrons and bonds.

7. RESONANCE: When Lewis Dot structures fail to accurately reflect reality, the next level is to use RESONANCE: draw multiple reasonable Lewis Dot structures and the real molecule will resemble the hybrid of these RESONANCE STRUCTURES.

3.   Multiple Bonds, Orbitals, Hybridization

1. Multiple Bonds. As we saw previously for carbonate, sometimes two atoms will share more than one pair of electrons. This is called a multiple bond. For carbon, these can be DOUBLE (2 pairs of electrons shared) bonds or TRIPLE (3 pairs of electrons shared) bonds. From a VSEPR perspective, a carbon with all single bonds has a TETRAHEDRAL shape, a carbon with a double bond and two single bonds has a TRIGONAL PLANAR shape, and a carbon with either a triple and a single or with two double bonds has a LINEAR shape. The following pictures shows Lewis Dot Structures and VSEPR for multiple bonds.

linear   trigonal planar

2. Formal Charge. One final thing with Lewis Dot structures. One helpful thing they allow us to do is assign "Formal Charge". In the carbonate ion structures we showed before, there was a charge on two of the oxygen atoms. How did we know? In the Lewis Dot structure, we add up the number of electrons a certain atom has, and compare with the number of protons it has. If it has more protons – one (+) charge for each. More electrons – one (–) charge for each. Core electrons and electrons in lone pairs count entirely for the atom that has them, electrons in shared pairs (bonds) count as ½ for each atom of the bond. So in the Carbonate ion $(CO_3)^{2-}$ drawn above for resonance: Carbon has 4 shared pairs, 8 electrons times ½ = 4 electrons, plus the 2 inner shell electrons gives 4 +2 = 6 electrons. Carbon also has 6 protons, so the Formal Charge is 0. For the double bonded Oxygen: 4 lone pair electrons = 4; 4 double bond electrons 4 x ½ = 2; and 2 inner shell electrons so 4+2+2 = 8 electrons. Oxygen has 8 protons, so the Formal Charge is 0. For the single bond Oxygens: 6 lone pair electrons = 6; 2 single bond electrons 2 x ½ = 1; and 2 inner core electrons so 6 + 1 + 2 = 9 compared to 8 protons gives a Formal Charge of -1. You will likely be expected to calculate formal charge a few times in a typical Organic class.

3. Orbitals: *s* and *p*. In our simple description of atoms, the protons and neutrons are in the center or nucleus, and the electrons on the outside. But there is more detail to how the electrons are arranged on the outside. We have already mentioned that the electrons are arranged in layers or shells. But there is even more detail that will help us. This next section is a little tedious, but understanding it will help you understand other things later. Electrons would like to be at the lowest energy possible. But they can't have just any energies – only certain amounts of energy are allowed. (Don't worry about why.) Amounts are "quantities" so they are called "quanta" (singular = quantum) and the study of how physical systems populate the allowed quanta is called "quantum mechanics". We will call the amounts / quanta "energy levels". Furthermore, for a given energy level, only 2 electrons are allowed to have it! Once 2 electrons have that energy – the next electron has to go to a higher

energy level. The energy levels correspond to areas in space around the nucleus of the atom. Electrons are light and fast and can (in theory) be just about anywhere in the universe – but they spend 90% of the time close to their nucleus. The shape of the volume where they spend 90% of their time is called an "orbital" (from the old idea that electrons orbit the nucleus like planets around the sun, since different planetary orbits also are different energy levels, and planetary motion is "celestial mechanics"). Each energy level / orbital only holds two electrons – these are connected in a special way through magnetism ("paired"), so they are called a "pair". The first shell has only one energy level / orbital. This orbital has a spherical shape – and it is called an "s" orbital: the 1s. These are the valence shell electrons Hydrogen uses to bond, but are the inner core electrons for Carbon, Nitrogen, Oxygen and Fluorine. For C, N, O and F, the valence shell electrons are in the second shell. This shell has a spherical orbital as well - the 2s. Remember that this shell holds 8 electrons (the octet), so there need to be three more orbitals. There are two more requirements for electrons and orbitals: first, in describing the orbitals we treat electrons as waves, so there can be (+) and (-) amplitude or "phase"; second, all the orbitals in an atom have to "overlap" to zero. Overlap is the sum of the product for two orbitals: where the two orbitals are both (+) or both (-) the product is (+); where one is (+) and one is (-) the product is (-). So the amount of (+)/(+) and (-)/(-) areas always have to equal the amount of (+)/(-) areas. In order to fulfil these requirements, the three other orbitals have a "dumbbell" shape and are called "p" orbitals – 2p. There are three, and they are all at 90° angles – one along the x axis, one along the y axis, one along the z axis. One side of the dumbbell is (+) and one is (-), and this makes the zero overlap requirement work among the 1s, 2s, and 2p orbitals! (The 2s is actually one sphere inside another, one (+) and one (-).)

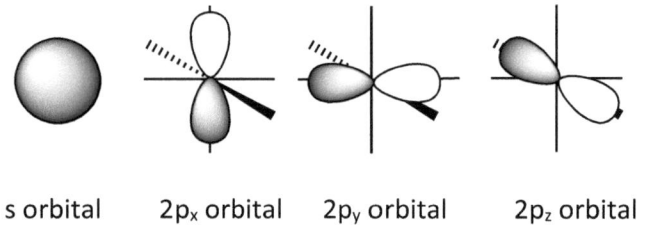

s orbital   2p$_x$ orbital   2p$_y$ orbital   2p$_z$ orbital

4. Covalent bonding in H$_2$, Using Molecular Orbitals. We can now use these orbitals to better understand covalent bonds. Hydrogen has a single electron in the 1s orbital. When 2 hydrogen atoms come together, the two hydrogen 1s orbitals mix together by simple addition. These two atomic orbitals (AO's) now make two molecular orbitals (MO's) – energy levels that exist when the two nuclei are sufficiently close together. These are σ = 1s + 1s and σ* = 1s – 1s. In the σ orbital, the electrons can spread out over both atoms, so the energy is lower – and this holds the atoms together = the covalent bond. So σ is called a bonding orbital. On the other hand, σ* has a higher energy than the contributing s orbitals – the electrons have to change phase and this always increases their kinetic energy. So σ* is called an anti-bonding orbital. Since the two hydrogens each bring one electron, these two electrons can both go into the bonding σ orbital to fill it up – no electrons go into the σ* orbital. This makes the bond between the two hydrogens: staying close together allows the electrons to have lower energy in the σ molecular orbital instead of higher energy in the individual 1s atomic orbitals.

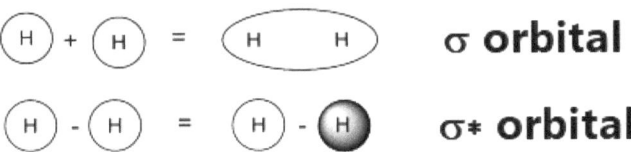

σ **orbital**

σ* **orbital**

Addition and subtraction of Hydrogen 1s orbitals

5. Can we use this simple approach with methane, CH_4? We can make a bond using the carbon 2s and a hydrogen 1s that is just like the bond between two hydrogens. Then we can make bonds from the 3 carbon 2p orbitals with 3 hydrogens, using their 1s orbitals. What will this look like? The hydrogens attached to the p orbitals make 90° angles, and the hydrogen attached to the carbon s orbital would be as far from these as possible – about 135°. But that isn't what VSEPR predicted, and it isn't what we find when we measure the bond angles of methane – the real methane has the ~109° bond angles VSEPR predicts. So this simple orbital to orbital approach fails to predict reality. What works best is to make orbitals by adding and subtraction combinations of all the carbon orbitals (2s and 2p) and the four hydrogen orbitals. But that is very complicated (working out the exact combinations that give the best bonding takes a lot of higher math and some computing power – it can't be solved by hand), and gets even more complicated as the molecules get bigger - so there is an intermediate approach called "hybridization". Hybridization is not entirely accurate in describing organic molecules, but your organic chemistry professor will expect you to know how to do it! So let's figure it out.

6. Here is how we hybridize for methane. Before making bonds with the carbon orbitals, we make new orbitals that are combinations of them: $2s+2p_x+2p_y+2p_z$, and 3 others with different mixtures of adding and subtracting (like $2s + 2p_x - 2p_y + 2p_z$) so that we make 4 new orbitals called sp^3 hybrids, because they are made from the s and all 3 of the p's. The 4 unique combinations make orbitals that have a small lobe on one side of the nucleus and a larger lobe on the other – kind of like a p orbital, but swollen on one side and shrunk on the other. These sp^3 orbitals point to the corners of a tetrahedron, as VSEPR predicts and experiments confirm. We put one of carbons 4 valence electrons in each sp^3 orbital. So to make methane, each of the sp^3 orbitals combines with a hydrogen 1s to make a bonding orbital and antibonding orbital as we did for H-H. The electrons (one from C sp^3, one from H 1s) go into the bonding σ orbital. In our hybridization model, whenever we have a carbon with four single bonds, that carbon will be sp^3 hybridized. The sp^3 hybrid orbitals can make bonds to H, C, O, and so on.

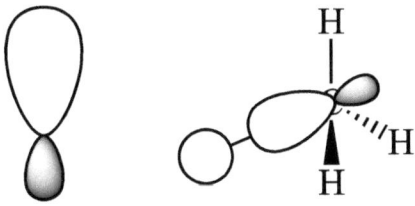

An sp^3 orbital, and methane showing one of the bonds as sp^3 + 1s

7. If carbon makes a double bond, it has to hybridize differently. The 2s orbital will mix with two of the 2p orbitals. These make three orbitals called sp^2 – made from one s and two p's. Like the sp^3, these have a small lobe on one side of the nucleus and a large lobe on the other side. If we use the $2p_x$ and $2p_y$, the resulting orbitals will be in the xy plane, pointing at the corners of an equilateral triangle at 120° angles – the trigonal planar arrangement predicted by VSEPR! The $2p_z$ orbital remains unchanged. The double bond is formed between two of these sp^2 hybridized carbons. They combine their sp^2 orbitals to make a σ bond, while their $2p_z$ orbitals

combine side to side to make a π bond. The π bond resembles a p orbital – one lobe above and one below. The π bond cannot rotate – rotation would break the interaction between the two p orbitals. Any carbon that makes one double bond – whether to C, N or O – will be sp² hybridized. THAT you need to know.

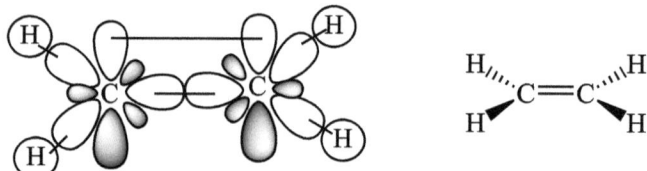

Ethene with its C=C double bond: cartoon showing orbitals of C's and H's and line structure

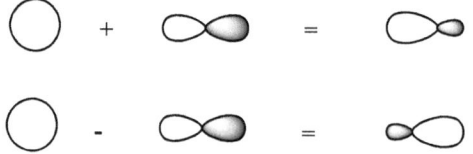

How the p orbitals add to make a π orbital

If carbon makes a triple bond, it has to hybridize in a third way. Only one of the 2p orbitals mixes with the 2s – we can use the $2p_x$. These are called sp hybrid orbitals, and they also have a small lobe to one side of the nucleus and a larger lobe on the other side. They point in opposite directions – 180° apart, the way VSEPR predicts. To make a triple bond, two sp hybridized carbons come together, make a σ bond with the sp hybrid orbitals, and two π bonds with the $2p_y$ and $2p_z$ orbitals. Any carbon that makes a triple bond is sp hybridized, and any carbon that makes two double bonds is also sp hybridized. Here we show how the sp hybrids are made by adding and subtracting s and p orbitals. These hybrids are the easiest to visualize how they are made, but the process is the same for sp² and sp³ – just a little more complicated since we are mixing more orbitals.

Making the two sp hybrids by adding and subtracting the s orbital and the $2p_x$ orbital

Quick Review.

1. MULTIPLE BONDS: Atoms can share multiple pairs of electrons. Sharing 2 pairs = DOUBLE BOND, sharing 3 pairs = TRIPLE BOND.

2. FORMAL CHARGE: In Lewis structures, we can calculate a FORMAL CHARGE. The Formal Charge is calculated from the number of electrons (all inner core electrons, all lone pairs, ½ of shared pairs) compared to the number of protons in the nucleus.

3. ORBITALS: Electrons in shells are organized into orbitals. Shell 1 has only one orbital, the 1s. Shell 2 has the 2s and 2p. The larger shells have more, but they are not important in organic chemistry.

4. MOLECULAR ORBITALS: We can understand bonding using orbitals. Hydrogen is the simplest and best example. Two hydrogens come together, their 1s orbitals combine into bonding σ and antibonding σ^*. Their electrons go into the bonding σ, forming the bond.

5. WHY HYBRIDIZATION: This simple approach fails for connecting carbon to four hydrogens in methane, predicting the wrong bond angles.

6. sp^3 HYBRIDIZATION: We combine the 2s and the three 2p's to make the sp^3 hybrids, these then make the correct bond angles.

7. sp^2 and sp HYBRIDIZATION: To make double bonds, we combine the 2s and two 2p's to make sp^2 hybrids, leaving one p orbital to make the second bond. To make triple bonds, we combine the 2s and one 2p to make sp hybrids, leaving two p orbitals to make the second and third bonds.

4.   IUPAC Nomenclature

1. The IUPAC is the International Union of Pure and Applied Chemistry. They are important to us because they establish the rules for naming chemical compounds. Because there are millions of compounds we need a SYSTEMATIC way of naming them. The rules are designed to create a one-to-one correspondence between name and structure: If you know the name you can follow the rules and draw the structure; if you have the structure you can follow the rules and generate the name. This is very important for organic chemistry. It will pay dividends to know the rules and be able to name compounds. Most professors will have 10-20% nomenclature on most exams – and those are easy points you do NOT want to miss! There is nothing difficult about the nomenclature rules – you just have to put in the time to learn how they work.

2. The nomenclature for organic compounds begins with compounds that are straight chains of carbon atoms with the rest of their valence positions taken up by hydrogens. Like this: $CH_3-CH_2-CH_2-CH_3$. Since they are made up of just hydrogen and carbon, these molecules are called HYDROCARBONS. These kinds of hydrocarbons, with all single bonds between the carbons, are called ALKANES. The ones that are straight chains are LINEAR ALKANES. *The names of the linear alkanes are the starting point for all other names.* They are called the ROOT NAMES or PARENT NAMES. All the other names are then made by adding prefixes, suffixes, and so on. So it is very important to know the names of the linear alkanes – they are the foundation everything else is built on!

3. In sophomore level organic chemistry, we usually limit the names to chains up to ten carbons long. There are longer chains, of course, and the IUPAC has names to cover chains that are over one hundred carbons long. But ten is sufficient for most purposes. So here they are:

Chain Length	Structure	Name
1	CH_4	Methane
2	CH_3-CH_3	Ethane
3	$CH_3-CH_2-CH_3$	Propane
4	$CH_3-CH_2-CH_2-CH_3$	Butane
5	$CH_3-CH_2-CH_2-CH_2-CH_3$	Pentane
6	$CH_3-CH_2-CH_2-CH_2-CH_2-CH_3$	Hexane
7	$CH_3-CH_2-CH_2-CH_2-CH_2-CH_2-CH_3$	Heptane
8	$CH_3-CH_2-CH_2-CH_2-CH_2-CH_2-CH_2-CH_3$	Octane
9	$CH_3-CH_2-CH_2-CH_2-CH_2-CH_2-CH_2-CH_2-CH_3$	Nonane
10	$CH_3-CH_2-CH_2-CH_2-CH_2-CH_2-CH_2-CH_2-CH_2-CH_3$	Decane

The first four – methane, ethane, propane, butane – are historical, so you just have to memorize them. The others are based on the length of the chain – pentane for 5, hexane for 6 and so on. If you know other names with these roots – like pentagon, hexagon, heptathlon – these are easier to remember. But – DO LEARN THEM!

4. But not all alkanes are linear. Most alkanes are BRANCHED ALKANES. That means that other carbon chains come off of the main chain. The way we will name these – detailed rules to follow – is to find the *longest* chain, call that the *main* chain and name it with the linear alkane ROOTNAME for that length of chain. Then the smaller chains that come off of the main chain are named, and located. To locate them we NUMBER all the carbons in the main chain – there are rules for the numbering. We name them as ALKYL groups or "substituents". (Since these "substitute" for hydrogen in the parent linear alkanes, they are called "substituents".) The names of the alkyl groups are based on the linear alkanes. We remove the –ANE ending, and replace it with the ending –YL. This ending means "substituent group". So the names correspond to the alkanes: methyl for $-CH_3$, ethyl for $-CH_2-CH_3$, propyl for $-CH_2-CH_2-CH_3$, butyl for $-CH_2-CH_2-CH_2-CH_3$ and so on – pentyl, hexyl, heptyl, octyl, nonyl, decyl. The perceptive student may ask "If there can be branches off the main chain, can there be branches off the smaller chains?" Of course there can, and branches off the branches off the branches – *ad infinitum*, as the philosophers used to say. BUT – we won't encounter much of that. There are two ways to handle it. The old way was to generate variant names for the branched branches. The newer IUPAC way is to treat the branched branches like the main chain – give the branch its own numbers (distinguished by "primes" 1' vs. 1) and use a similar system to the one we will detail momentarily. The old system was OK for small, 3-5 carbon branched branches, but quickly became unmanageable. But the names for those 3-5 carbon branched branches have stuck around – and will be described below.

5. Knowing the names of the groups, we can now proceed to show how to name ANY alkane. This is done by following a sequence of "rules". First – be sure you are dealing with an ALKANE – all carbons and hydrogens, no double bonds, no triple bonds.

RULE 1. Find the longest continuous chain of carbon atoms. Assign this the name of the linear alkane of the same length. This will be the ROOT NAME. This sounds simple – but BE CAREFUL! In most examples, the longest chain is drawn left to right, the way we read and write. But it doesn't need to be drawn this way – a tricky professor may have the longest chain make left turns and right turns so it is harder to see!

RULE 2. Locate and identify all branches – assign them their names (methyl, ethyl, etc.). If there are NO BRANCHES, then it is a LINEAR ALKANE and the ROOT NAME is the name!

RULE 3. Number the chain. We have to choose one end to number from, and this can be a little tricky. The goal is to have lower numbers over higher numbers, but it would be time-consuming in the case of lots of substituents to add all the numbers up. So the master rule for numbering is to number so as to have the "lower number at the first point of difference". What always works is to compare carbons from each end. How many substituents are there on each? Zero? One? Two? When one end has a substituent where the other doesn't – that is the FIRST POINT OF DIFFERENCE. To give a lower number – number from the end closer to the point of difference. In some cases, it may be so obvious that you don't need to compare carbons one at a time – a 6 carbon chain, for example, with a single substituent, one carbon in from the left end. With no other substituents, the left end is the obvious place to start the numbering (number from the left – substituent on carbon 2, from the right – substituent on carbon 5). But what if there is no point of difference? The substituents are the same from either end! Then – IT DOES NOT MATTER WHICH END YOU NUMBER FROM! So pick an end - number from the left. Be sure and GIVE EVERY CARBON IN THE MAIN CHAIN A NUMBER.

CH₃ structures:

$$CH_3$$
$$H_3C-CH-CH_2-CH_2-CH_2-CH_3$$

———→ number this way

not this way ←———

2-methylhexane

$$CH_3 \quad CH_3$$
$$H_3C-CH_2-C-CH_2-CH-CH_3$$
$$CH_3$$

number this way ←———

———→ not this way

2,4,4-trimethylhexane

RULE 4. The name of the branched alkane is given by listing each substituent with its number before the ROOT NAME. Numbers and letters are always separated by hyphens; any place where two numbers come together, the numbers are separated by commas. The last substituent name runs right into the root name. So in the case described above – methyl group on carbon 2 of a 6 carbon chain – the name is 2-methylhexane. When there are more than one of a given type of group, these are all combined together and prefixes are used to indicate how many there are: di for 2, tri for 3, tetra for 4, penta for 5, hexa for 6 and so on. So if there were 2 methyl groups on carbon 2 of a 6 carbon chain: 2,2-dimethylhexane. Note that we need 2 numbers to tell where the 2 substituents are – it is a common error for students to just call this 2-dimethylhexane – but since the second methyl could be on carbon 3 or 4 or 5, we need to be able to tell apart the molecules 2,2-dimethylhexane from 2,3-dimethylhexane and so on. Here also we see the comma between two numbers.

RULE 5. When listing different types of substituents before the root name, they are listed in ALPHABETICAL order. So for the common substituents: butyl, then ethyl, then methyl, then propyl. All the other rules are essential for our purpose of having each unique structure have a unique name. But this one is not – we could list the substituents by size or by number and the names would still correspond. (Like listing people's names as Jane Doe or Doe, Jane.) So why alphabetical? Chemists would probably choose size – I blame the librarians and database managers. Our alphabetization is a little unusual. Any prefixes (di-, tri-, iso- etc.) don't count. Butyl comes before ethyl, and so does tributyl! Propyl is after methyl and so is isopropyl!

6. Cycloalkane nomenclature. Not all carbon molecules are chains. In some cases, a chain of carbons connects the two ends to make a ring. There are then a few new wrinkles in nomenclature. To name a ring of carbons, we count the carbons in the ring. We use the same names as the linear alkanes, but add the prefix "cyclo-". The smallest ring is 3 carbons, and we call it cyclopropane. A 4 carbon ring is cyclobutane, a 5 carbon ring is cyclopentane, and so on. If there are substituents on the ring, we still want to identify by name and number. But since the ring has no ENDS, the numbering is a little different. If there is only one substituent, the carbon where it attaches is given the number 1. But when we name this compound, we don't need the 1, since with one substituent, it is ALWAYS on carbon 1. So a methyl group attached to a 6 carbon ring is called methylcyclohexane (technically it is OK to call it 1-methylcyclohexane, but don't! You sound like a newbie!) If there are two or more groups attached to the ring, the numbering is done so as to give lower numbers at the first point of difference again. So methyl groups on adjacent carbons of a six membered ring – we pick one to be carbon 1 then go around the ring so the next one is 2, not 6: 1,2-dimethylcyclohexane (NOT 1,6-dimethylcyclohexane). There is also a 1,3-dimethylcyclohexane (never 1,5-dimethylcyclohexane) and a 1,4-dimethylcyclohexane (here the numbers are the same whichever way we number).

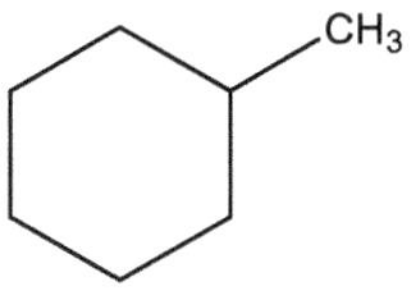

methylcyclohexane 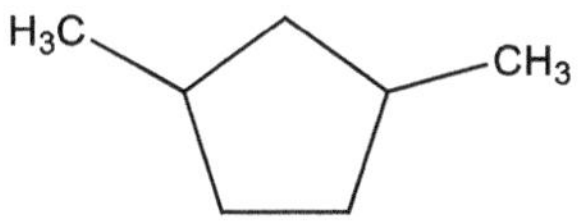   1,3-dimethylcyclopentane

7. IUPAC is a newer system – compounds existed and were named before it was devised. Sometimes these older names have survived through common use. We see these in the branched branch names given below, and we will see it again occasionally as we go through the nomenclature of various types of compounds. One method that survives is to identify types of carbons based on the number of carbons attached. If a carbon has one carbon attached, it is said to be PRIMARY and given the symbol of 1°. If a carbon has 2 carbons attached, it is SECONDARY and given the symbol 2°. If a carbon has 3 carbons attached it is TERTIARY and given the symbol 3°. If a carbon has 4 carbons attached it is QUATERNARY and given the symbol 4°. A carbon with 0 carbons attached is METHYL. This will also be important as the types of carbons can be different in their reactivity. These designations are also used in the old names that are still used for branched groups. We should learn the 1 branched 3 carbon group (isopropyl) and the 3 branched 4 carbon groups (isobutyl, sec-butyl, and tert-butyl). I will throw in one of the many branched 5 carbon groups (isopentyl) – it shows us how any "iso" group looks. Note how isopropyl, isobutyl, and isopentyl all have a chain one shorter than the parent chain: propyl is 3, so isopropyl is 2 carbons long, with the 3rd as a branch on the carbon one in from the end – isobutyl (butyl is 4) is 3 carbons long with the 4th as a branch on the carbon one in from the end, and isopentyl follows the same pattern. Sec-butyl connects through the secondary (CH) carbon, while tert-butyl connects through a tertiary (C) carbon. When we want to represent a general carbon (alkane or alkyl) structure, we represent it by the symbol "R".

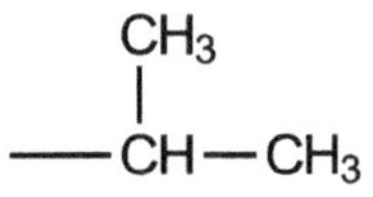

isopropyl 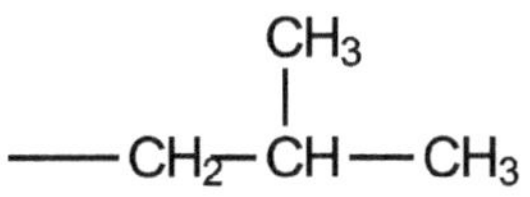   isobutyl 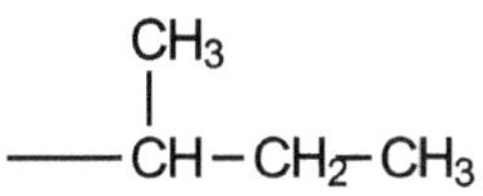   sec-butyl 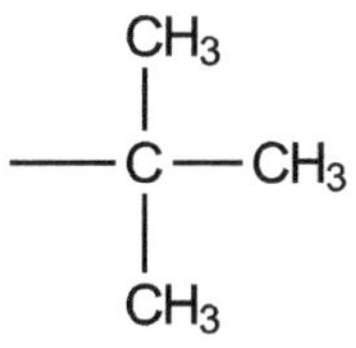   tert-butyl

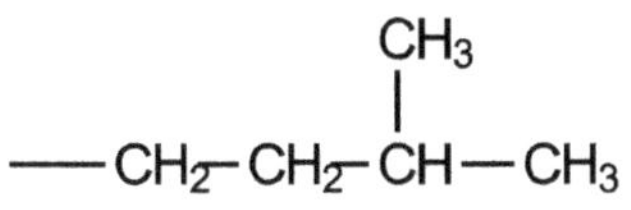

isopentyl

Quick Review

1. IUPAC: IUPAC is an organization of chemists – one thing they do is establish rules for nomenclature. The naming system is designed to generate unique names for each unique structure.

2. LINEAR ALKANES: The basis for all names in organic chemistry starts with the linear alkanes.

3. NAMES OF LINEAR ALKANES: Linear alkanes 1-10: Methane, Ethane, Propane, Butane, Pentane, Hexane, Heptane, Octane, Nonane, Decane.

4. BRANCHED ALKANES: Branched alkanes are named as linear alkanes with alkyl substituents. Substituent names come from the linear names but with the ending –yl: methyl, ethyl, propyl, butyl. Some branched substituents are special: isopropyl, isobutyl, sec-butyl, and tert-butyl.

5. RULES of ALKANE NOMENCLATURE: 5 Rules for naming alkanes: (1) Identify Parent chain (2) Identify substituents. (3) Number the parent chain. (4) Substituents by name and number: 2-methylpropane (5) List substituents alphabetically.

6. CYCLOALKANES: Rings are named as cycloalkanes: cyclopropane, etc.

7. BRANCHED BRANCHES and more: Primary – secondary - tertiary – quaternary from the carbons surrounding a carbon, old names of the simple branched branches.

II. Alkanes, Conformations, Stereochemistry
5.   Alkanes: the Null Functional Group

1. You just learned the nomenclature of the Alkanes. Alkanes are the simplest of the hydrocarbons – compounds made of only hydrogen and carbon, with only single bonds between them. A common way to organize our knowledge in Organic Chemistry is through the use of FUNCTIONAL GROUPS. A functional group is a set of atoms with a certain bonding arrangement, such as the hydroxyl group: an oxygen and a hydrogen –O-H connected by single bonds. This functional group defines the compounds we call alcohols. These functional groups then have similar chemical and physical properties when attached to different hydrocarbon skeletons. Alkanes are the parent compounds of all organic molecules – they are the compounds with NO FUNCTIONAL GROUP. The table below shows some common functional groups and the types of compounds they define. As we learn organic chemistry, one way we organize information is by types of compounds. This is useful because it allows us to learn a common set of properties – chemical and physical – for each type of compound. So all alcohols have similar properties. The list in the Table is not exhaustive – there are numerous other functional groups – but covers the most common functional groups. When you have covered functional groups as a general concept, your professor may ask you to identify them.

Alkanes have NO FUNCTIONAL GROUP – they are sometimes represented as R-H.

Type of Compound	Functional Group Structure	Functional Group Name	Generic Structure
Alkene	C=C	Double bond	
Alkyne	C≡C	Triple bond	
Aromatic		Aromatic Ring	
Haloalkanes	-X (X = F, Cl, Br, I)	Fluoro, Chloro, Bromo, Iodo	R-X
Alcohols	-OH	Hydroxyl	R-OH
Ethers	-O-	Alkoxy	R-O-R'
Amines	-NH₂, -NH-, -N ⟨	Amino	R-NH₂
Aldehydes	-C=O + -H	Formyl	R-CHO
Ketones	-C=O	Acyl / Alkanoyl	R-CO-R
Carboxylic Acids	-C=O + -OH	Carboxyl	R-CO₂H
Esters	-C=O + -OR[	Carboxylate	R-CO₂R'
Amides	-C=O + -NH₂, -NHR, NR₂	Carboxamide	R-CONH₂
Acid Chlorides	-C=O + -Cl	Carbonyl Chloride	R-COCl
Nitriles	-C≡N	Cyano	R-CN
Thiols	-SH	Sulfhydryl	R-SH

2. We are mainly talking in this section about the alkanes. All alkanes have a general formula of C_nH_{2n+2}. The reason for this is apparent when you look at the structures of the linear alkanes above. In the middle of the linear

alkanes, all the carbons are CH_2 groups (to impress your professor call them "methylene" groups). But the two at the end are CH_3 (methyl) – think of these ends as CH_2-H instead. So all the carbons are CH_2, but the two ends are capped with 2 additional H's : C_nH_{2n} + 2 more H's = C_nH_{2n+2}. If we replace an H with a substituent like a methyl – we lose an H and get CH_3 – or CH_2-H. So each carbon group comes in as a C_nH_{2n} and the "original" H is at the end of the side chain, and the C_nH_{2n+2} ratio remains. This is as many H's as we can put on this number of C's. So we say the chain is "saturated" with H. So alkane type chains are called "saturated" hydrocarbons. In contrast, the alkenes and alkynes have fewer H's to form the double and triple bonds. They are not saturated – there is room for more H's - so they are called "unsaturated" hydrocarbons. You may see these on your food labels: fats and oils with all alkane structures are called "saturated fats" (oooh – BAD in the traditional dietary model – but research is ongoing) while those with alkene structures are "unsaturated fats" (GOOD in the traditional dietary model). It is good to know the general formula of the alkanes and how hydrocarbons are divided into saturated and unsaturated.

3. What are alkanes like? How do their melting points and boiling points compare to other compounds? Do they dissolve in water? We call these data points PHYSICAL PROPERTIES. Each type of compound you meet will have physical properties to learn. These are of modest importance – focus on nomenclature and chemical properties (reactions) first. Alkanes are all C-C and C-H bonds, which are nonpolar, so alkanes are NONPOLAR. Compounds that are NONPOLAR only have weak attractions to each other (known as LONDON forces, DISPERSION forces, or VAN DER WAALS forces – different names for the same thing). This decreases the tendency toward liquid or solid. The result: LOW MELTING POINTS and LOW BOILING POINTS. This will be the trend you see – more NONPOLAR ⇨ LOWER MP and BP; more POLAR ⇨ HIGHER MP and BP. As an example, nonpolar methane (mol. wt. = 16) and very polar water (mol. wt. = 18) can be compared: methane: MP= -182 ^{o}C, BP = -161 ^{o}C; water MP = 0° C, BP = 100 ^{o}C. Methane has low MP and BP, water has much higher MP and BP. As the molecular weight of alkanes increases, the MP and BP increase as well. Methane through butane are gases at room temperature (25 ^{o}C), pentane through ~C_{20} are pourable liquids. The linear alkanes don't become true crystalline solids at higher MW – they become highly viscous liquids (think motor oil), then gums, then waxes. At a sufficiently high MW they become "plastics" – polyethylene is ubiquitous in modern life, and is essentially a linear alkane with 100s or 1000s of CH_2 groups. The same weak forces keep the chains in the liquid and solid forms fairly far apart – so alkanes have low density (oil floats on top of water), usually about 0.65-0.75 g/mL (compared to 1.00 g/mL for water). The density of alkanes increases with MW. The old adage in chemistry is that "like dissolves like" so the very nonpolar alkanes and polar water do not mix (oil and water, you know) and the solubility of the alkanes is very low in water. Because we want a yes/no answer about solubility, anything that dissolves less than 10 g in 100 mL of water is "not soluble". The alkanes (typical 0.005 g in 100 mL) are not even close. The three properties we will usually note for classes of compound are BP, MP, and solubility. Others can be measured – density (low for the alkanes, about 0.7 g/mL compared to water at 1.0 g/mL), refractive index (how light bends going into the substance from air), color, odor, etc. We will occasionally mention the odor of certain classes of compounds – most alkanes do not have a notably strong odor. For the alkanes we do want to point out a few other things about MP and BP. Branches are known to lower the boiling point – more branches reduces the surface area, which reduces the intermolecular attraction. Branching often lowers the melting point as well, but this is less predictable as the MP also depends on molecular symmetry – more symmetrical structures have higher MP – and branching can increase symmetry.

How do physical properties show up on exams and quizzes? Usually as comparison questions : here are 2 compounds, which has the lower MP? This requires the two compounds to be different in a way that correlates with a physical property. So given 2 alkanes of equal MW, but one linear and one branched, which has the lower BP? Answer - The branched one!

ALKANE PHYSICAL PROPERTIES

Polarity: NONPOLAR   Refractive Index: 1.35 – 1.45
MP/BP: Low   Odor: Little to no odor
H_2O sol.: No   Color: None
Density: Low, ~0.7 g/mL

4. An important concept of organic chemistry is that of ISOMERS. As we examine the alkanes, we see isomers for the first time. ISOMERS are compounds with the same molecular formula but different structures. So for the molecular formulas, CH_4, C_2H_6, and C_3H_8 – there is only one way to connect the atoms: methane, ethane, propane. BUT for C_4H_{10}, we can either have all 4 carbons in a row (butane) or have 3 with a 1 carbon branch off the 2nd carbon (2-methylpropane). Obviously, these are different compounds. The difference is in the way the atoms are connected together or CONNECTIVITY. When molecules have the same molecular formula but differ in connectivity we call them CONSTITUTIONAL ISOMERS. Even here there are varieties – the difference in connectivity could be between two alkanes (same functional group) or between, for example, an alcohol and an ether (with different functional groups). These are called "positional isomers" when the functional group is the same and "functional group isomers" when different. This distinction is useful – positional isomers will have similar properties since they have the same functional group, but functional group isomers can have very different properties - but this isn't used as much.

$$CH_3-CH_2-CH_2-CH_3$$
Butane

$$CH_3-\overset{\overset{\displaystyle CH_3}{|}}{CH}-CH_3$$
2-Methylpropane
"isobutane"

Two additional things about isomers. First, there is a connection between isomers and nomenclature – if the structure is different, the name should be different. And the IUPAC nomenclature is set up to give each different structure a different name – even if it is only a difference in number: 2-methylpentane vs. 3-methylpentane. If you are unsure if two compounds are isomers or identical, name them. Same name = same structure! Second, the number of isomers grows enormously with size. As we saw, there is only 1 structure (isomer) for CH_4, C_2H_6, and C_3H_8, 2 for C_4H_{10}, 3 for C_5H_{12}. The number of isomers then grows very quickly:

Molecular Formula	Number of Isomers
C_4H_{10}	2
C_5H_{12}	3
C_6H_{14}	5
C_7H_{16}	9
C_8H_{18}	18
$C_{10}H_{22}$	75
$C_{20}H_{42}$	366,319
$C_{32}H_{66}$	27,711, 253,769

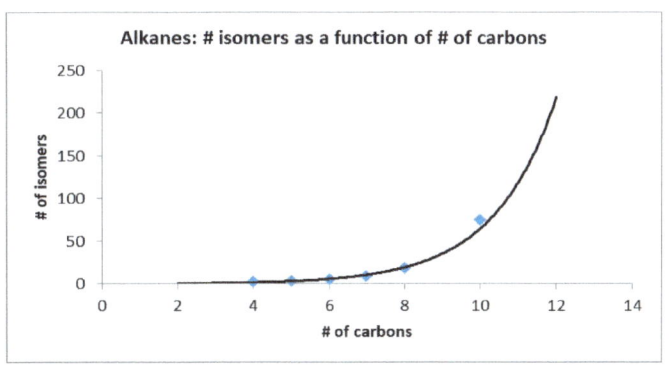

25

5. Alkanes are relatively unreactive. An old name for alkanes is "paraffin" – for example, if your supermarket stocks supplies for home canning, you may find "paraffin wax" – alkanes of high enough MW to have a waxy consistency. In Latin, "paraffin" means "no affinity" = not reactive. BUT there are a few reactions of alkanes. There are two important enough to be treated in sophomore Organic Chemistry. We will mention both here, but save the details of one for later. The two reactions are COMBUSTION and FREE RADICAL HALOGENATION.

COMBUSTION is oxidation of alkanes with molecular oxygen (O_2), usually from air. Chemically, this is a very boring reaction. Economically, it is crucial! Here is combustion of the four smallest alkanes, methane to butane:

$$CH_4 + 2\ O_2 \longrightarrow CO_2 + 2\ H_2O$$

$$C_2H_6 + 7/2\ O_2 \longrightarrow 2\ CO_2 + 3\ H_2O$$

$$C_3H_8 + 5\ O_2 \longrightarrow 3\ CO_2 + 4\ H_2O$$

$$C_4H_{10} + 13/2\ O_2 \longrightarrow 4\ CO_2 + 5\ H_2O$$

Note that the products are always the same: CO_2 and H_2O. That is why it is boring chemically – all that is different is balancing the equation! BUT the reactions release A LOT of energy – so it is important economically. You probably heat your house with natural gas (= methane) and power your car with gasoline (= blend of alkanes in the $C_6 - C_{11}$ range). All fossil fuels (gas, oil, coal) are hydrocarbons – if we did not have these to burn – welcome to the year 1750! (when coal first started being widely used – before that, you chopped wood to heat your house – better have a supply laid in for winter! Only horses for travel on land, sailing ships on water – America to Europe took months – no airplanes!)

FREE RADICAL HALOGENATION is a process where the H atoms on alkanes can be replaced with halogens – in theory F, Cl, Br, or I. In practice, only Cl and Br are used. If you mix alkanes and F_2 – RUN FOR YOUR LIVES! – they explode! On the other hand, iodine (I_2) does not react efficiently. Here is chlorination and bromination of methane. The details will be shown later.

$$CH_4 + Cl_2 \xrightarrow{\text{heat or light}} CH_3Cl + HCl$$

$$CH_4 + Br_2 \xrightarrow{\text{heat or light}} CH_3Br + HBr$$

6. As we saw before, alkanes can have rings of carbons, not just straight chains and branched chains. These are the cycloalkanes. They are very similar to the alkanes in most ways. They are nonpolar, low MP and BP, low H_2O solubility – all the physical properties are similar. Compared to a linear alkane with the same number of carbons, the cycloalkane will have a higher MP, higher BP (usually 10-20 °C), and higher density. This is because the connected ends allow greater contact between molecules, enhancing the attractive forces. So hexane has MP of -95 °C and BP of 69 °C while cyclohexane has a MP of 6 °C and BP of 81 °C. Cycloalkanes have a general formula of C_nH_{2n} – the two "extra" hydrogens at the end of the linear alkane chain are lost when the two ends of the chain connect to make the ring! The reactions of cycloalkanes are generally the same as those of alkanes – we will see that cyclopropane and cyclobutane can be a little unusual. Cycloalkanes are often represented by skeletal

structures in the form of the regular polygons – triangle, square, pentagon, hexagon. The ease of drawing the rings tends to make them a favorite of professors on tests and quizzes – so give them a little extra studying.

7. The single bonds of alkanes can generally rotate freely. Because molecules are small, they move fast and rotate fast. We can imagine all the methyl groups spinning like little fans – the rate of rotation is fast: like 10^{10} per second! But cycloalkanes can't. Because they are closed rings, to rotate freely, the bonds would have to pass through each other! So cycloalkanes can twist back and forth, but can't rotate. This creates a new isomer situation. Each carbon of the cycloalkane is a CH_2 group. If we imagine the ring is planar (they aren't always – more soon) and horizontal (for convenience), then one H is UP and one H is DOWN. If we replace an H with a substituent – say a methyl group – it has to go on one side of the ring (UP or DOWN). We can always flip the ring over to change up and down, so these are the same. BUT -if we replace a second H on a different carbon, suddenly we have two choices: replace the H on the SAME SIDE as the existing methyl, or replace the H on the OPPOSITE side. So there are two *different* 1,2-dimethylcyclopropanes! One has both methyl groups on the SAME SIDE, one has them on OPPOSITE SIDES. These compound have the same molecular formula and the same connectivity – but they are still DIFFERENT – they differ in the ARRANGEMENT IN SPACE of the atoms. They are a new type of isomer: STEREOISOMERS. We now need to be able to distinguish them – a new piece of the naming system is required! Two old German words were enlisted by the old German chemists who devised this part of the system: CIS means same side, TRANS means opposite side. So we call the two compounds *cis*-1,2-dimethylcyclopropane when the methyls are on the same side, and *trans*-1,2-dimethylcyclopropane when they are on opposite sides.

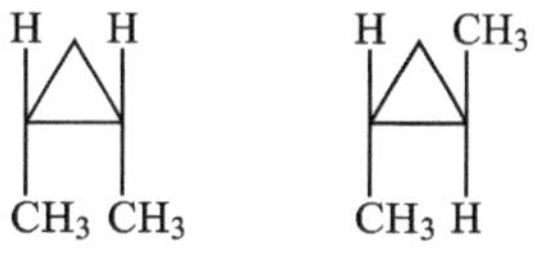

Stereoisomers of 1,2-dimethylcyclopropane: *cis* and *trans*

Quick Review

1. FUNCTIONAL GROUPS: We organize Organic Chemistry knowledge by functional groups. Know the common ones!

2 ALKANES: Alkanes are the compounds with just carbons and hydrogens connected by single bonds – no functional group! They are C_nH_{2n+2}, "saturated" with H, therefore saturated hydrocarbons.

3 PHYSICAL PROPERTIES: Alkane physical properties: nonpolar, low MP and BP, low H_2O solubility, low density. Branched alkanes have lower BP than straight chain alkanes.

4 ISOMERS: same molecular formula, different structure. If the difference is connectivity: constitutional isomers. The number of isomers of alkanes grows enormous with increasing length of chain.

5 ALKANE REACTIONS: combustion and free radical halogenation.

6 CYCLOALKANES: rings of carbons. They are a lot like alkanes, but have higher MP, BP and densities.

7 STEREOISOMERS: Substituted cycloalkanes can have stereoisomers – isomers that differ in the arrangement of atoms in space. Two substituents on the same side of the ring is *cis*, if on opposite sides, *trans*.

6.   Conformational Analysis

1. As described above, alkane C-C and C-H bonds can rotate freely. For the C-H bonds, this is trivial: just a little H atom spinning at the end of the bond. But for the C-C bonds, things are a little more complex. The changes in atomic arrangements in a molecule that result from rotation around single bonds are called CONFORMATIONS. The energy of the molecule can be different as the atomic arrangements change. The study of these energy changes is called CONFORMATIONAL ANALYSIS. While this is a very deep subject, we will only look at it very briefly.

2. The simplest molecule to examine is ethane. On each CH_3 group we chose one H to track the rotation – we will make one red and one green. We define a DIHEDRAL ANGLE - the angle between the two planes defined by the two H's. Three points define a plane, so the two planes are H-C_1-C_2 and H-C_2-C_1. When the two H's are lined up, the angle is 0 and we call the structure ECLIPSED. We show this with a special drawing called a NEWMAN PROJECTION. In this drawing, we look down the axis of the C-C bond, which is represented by a circle. The bonds to the near carbon (C_1) are shown as three lines that are in front of the circle and meet in the middle. The bonds to the far carbon (C_2) are shown as three lines that are behind the circle – so they start at the circle and point out. When we draw the eclipsed conformation, we offset the front and back bonds a little for visibility. As the C-C bond rotates, when the dihedral angle reaches 60°, the rear H is exactly between 2 of the front H's. This conformation is called STAGGERED. As the rotation continues, there is an eclipsed conformation at 120°, staggered at 180°, eclipsed at 240°, and staggered at 300°. At 360° = 0° - back to eclipsed. Analysis of the energy of the eclipsed and staggered conformations reveals that the eclipsed is higher in energy by a little bit – 12.6 kJ/mol (3.0 kcal/mol if your professor is old fashioned) to be exact. Since ethane just cycles between eclipsed and staggered conformations as it rotates, the energy vs. angle plot is a sine wave. Why is eclipsed higher in energy? The simple way to think about it is that the electrons in the C-H bonds repel each other, so it is lower energy for them to be as far apart as possible – staggered. This is a conformational analysis version of VSEPR! We can do an equilibrium treatment of the two states – with an energy difference of 12.5 kJ/mol, we can calculate the percentage of staggered vs. eclipsed (the equilibrium constant is 162, so about 99.4% staggered).

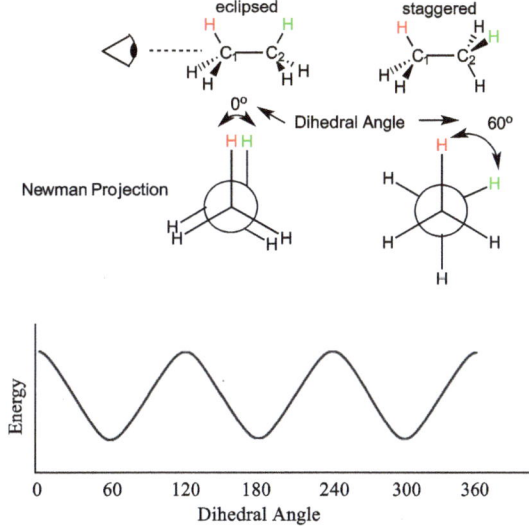

3. Ethane is very simple. The other alkane we usually analyze is butane. Here we look at the central C-C bond. The two carbons in question have two bonds to H and one to a CH_3. The CH_3 groups are the unique groups, so we

use them to define the dihedral angle in butane. When the two CH_3 groups are eclipsed (CC eclipsed), the dihedral angle is 0°, and this conformation has the highest energy, since the two CH_3 groups actually bump into each other a little. As we rotate, the back CH_3 moves between the front CH_3 and an H at 60°. This is a staggered-type conformation that is called GAUCHE, and the energy is lower. As rotation continues, the back CH_3 eclipses a front H at 120° (CH eclipsed), and the energy goes up – but not as much as the CC eclipsed. At 180°, another staggered-type conformation occurs with the two CH_3 groups directly across from each other, each between two hydrogens. This is the lowest energy conformation, called ANTI. As rotation continues, there is another CH eclipsed conformation at 240°, and another gauche conformation at 300°. We are back to CC eclipsed at 360°=0. The plot of energy vs. dihedral angle looks like the addition of two sine waves – with the peak at 0 and trough at 180, but additional oscillations between the gauche and CH eclipsed forms. The CC eclipsed is 20.9 kJ/mol (5.0 kcal/mol) above anti, CH eclipsed is 14.6 kJ/mol (3.5 kcal/mol) above anti, and gauche is 3.8 kJ/mol (0.9 kcal/mol) above anti. If you know the butane conformations, you should be able to apply them to almost any other alkane. An equilibrium treatment of butane shows that the anti/gauche ratio is 4.5:1 at room temperature.

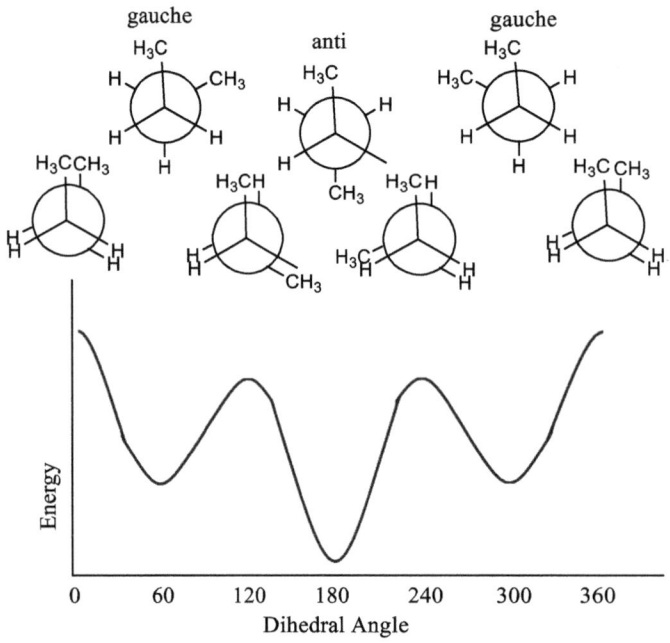

4. The increased energy of some conformations relative to others is called STRAIN. The lowest energy conformation is said to be "unstrained": this would be staggered for ethane, anti for butane. The strain due to rotation around single bonds as we see in ethane and butane is called TORSIONAL strain. (Rotational motions are called "torsions".) There are other sources of strain. If the bond angles deviate from the tetrahedral optimum of 109.5°, the energy goes up – this is ANGLE strain. If two pieces of a molecule come too close together, so the atoms are in each other's space (the "van der Waals radius" that is used to define the size of an atom) this is called "van der Waals" strain (or "non-bonded strain").

5. One place where we see different types of strain is the cycloalkanes. Cyclopropane has three carbons, and three points define a plane, so cyclopropane has to be planar (flat). An equilateral triangle has internal angles of 60°, which is considerably compressed from the tetrahedral 109.5°. This causes angle strain. Because it is planar, the bonds of cyclopropane are forced to be eclipsed. Therefore cyclopropane also has considerable torsional strain. With the combination of angle and torsional strain, cyclopropane is the most strained cycloalkane –

numbers later. If cyclobutane were a planar square, the angles would be 90°, and the C-H bonds would also be eclipsed. Since the actual cyclobutane can twist, two carbons move up and two move down. This actually makes the angle strain worse (~88°), but relieves some of the torsional strain. The shape is said to be "puckered". Planar cyclopentane would have bond angles of 108° – nearly unstrained – but would be fully eclipsed giving high torsional strain. So cyclopentane also twists – one carbon up more than the others. Like cyclobutane, this increases angle strain (angles are ~105°) but relieves torsional strain. The "up" carbon looks like the flap of an envelope, so this is called the "envelope" conformation. Cyclopentane only has a little strain, and is very flexible – which carbon is "up" changes rapidly. Planar cyclohexane would have 120° angles – this would be strained from having the angles too BIG! So cyclohexane twists, it moves one end down and the opposite end up. This is said to look like a chair – the "chair" conformation. After it has done this twist, the bond angles are 109.5° – unstrained. As we will see in more detail in the next section, the result also is staggered conformations – no torsional strain either! So cyclohexane is the unstrained ring.

cyclobutane   cyclopentane   cyclohexane

puckered   envelope   chair

While the concept of strain is not too hard to grasp – carbon has optimal bond angles and dihedral angles, and if a structure deviates the energy goes up – actually measuring values can be more difficult. Here is the usual way it is done. Combustion of alkanes (remember : alkane + oxygen gives carbon dioxide + water + heat) releases energy. If we measure the energy released from straight chain alkanes as they get longer, we find that once we get past the smallest alkanes, every added CH_2 adds the same amount of energy. This is then identified as the energy of an unstrained CH_2. If we then measure the energy released from a cycloalkane (all CH_2 groups), we can compare these numbers in two ways. The total energy released by the cycloalkane compared to the energy of an equal number of unstrained CH_2 groups gives (as the difference) the total strain energy, while the total energy released divided by the number of CH_2 groups can be compared to the unstrained CH_2 to find the strain energy per carbon (or per CH_2 group). The results are below. Note that the total strain energy of cyclobutane and cylopropane is similar, but the strain per CH_2 is quite a bit higher for cyclopropane. Cyclohexane is the unstrained ring, then strain increases for larger rings.

Ring	Total Strain kJ/mol (kcal/mol)	Strain per carbon kJ/mol (kcal/mol)
Cyclopropane	115.5 (27.6)	38.5 (9.2)
Cyclobutane	110.4 (26.4)	27.6 (6.6)
Cyclopentane	27.2 (6.5)	5.4 (1.3)
Cyclohexane	0 (0)	0 (0)
Cycloheptane	26.4 (6.3)	3.8 (0.9)
Cyclooctane	40.2 (9.6)	5.0 (1.2)

Key points: cyclopropane and cyclobutane are very strained, cyclopentane and cycloheptane mildly strained, cyclohexane unstrained.

6. Cyclohexane is the unstrained ring, which makes it a very common ring, and the most studied of all rings. As mentioned above, one carbon twists down, the opposite one twists up to make the chair conformation. When this happens, the C-H bonds rotate to either point OUT or UP/DOWN. The ones that point OUT are called EQUATORIAL (they are on the "equator" of the ring), while the ones that point UP or DOWN are called AXIAL (they point along the North/South axis). Each carbon has one equatorial and one axial C-H bond. If we take the cyclohexane ring and twist in the opposite direction – the carbon that was down goes up, and the carbon that was up goes down – all of the C-H bonds reverse: the axial become equatorial, the equatorial become axial! So there are two different chairs for cyclohexane – a left facing chair and a right facing chair. When cyclohexane moves from one chair to another, it is called a "chair flip" – but the ring itself doesn't flip over, the ends just flip up and down. There is also a conformation where both ends move up (or down) – the angles are 109.5° but all the bonds are eclipsed. This conformation is called the "boat" and is high energy. As the chairs change from one to the other, they go through other conformations called the "half-chair" and "twist-boat". These are higher energy than the chair, which makes the chair flip process require energy. At room temperature, the flipping is still very rapid. You should understand cyclohexane. It is useful to build models of cyclohexane that you can rotate around and see all of the conformations. Understand the axial and equatorial positions on each carbon.

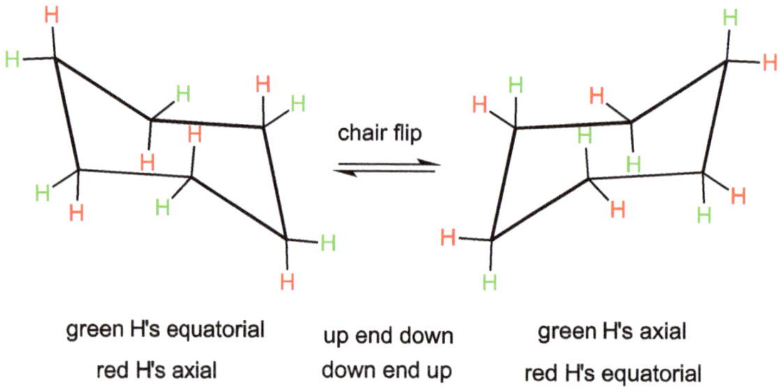

	chair flip	
green H's equatorial	up end down	green H's axial
red H's axial	down end up	red H's equatorial

boat   twist-boat

7. The two chairs of cyclohexane are equal in energy, so at equilibrium they exist in equal amounts. But what about methylcyclohexane? In one chair, the methyl group is axial, and in the other it is equatorial. In the equatorial position, the methyl group is anti to the ring. In the axial position, it is gauche to the ring. We remember from butane that gauche is higher energy than anti, so the chair where the methyl group is equatorial is lower in energy, by 7.1 kJ/mol (1.7 kcal/mol). Equilibrium calculations give the ratio of equatorial to axial as 95:5 for a methyl substituent.

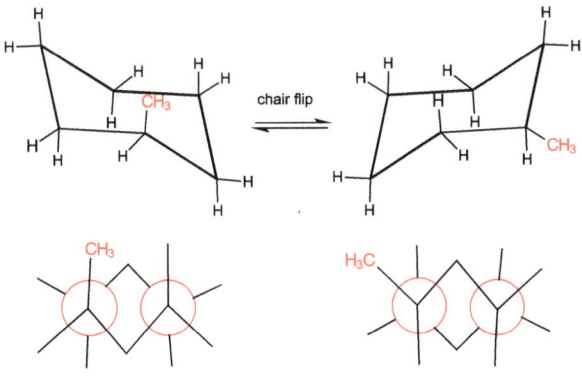

What about dimethylcyclohexanes? It pays to recognize a pattern in cyclohexane. If we start at one carbon (call it carbon 1) and the upper position is axial, and the lower equatorial; on the next carbon, the upper position will be equatorial and the lower axial. As we move around the ring, it continues to alternate. So positions 1, 3, and 5 will be one way (axial up or axial down) and positions 2, 4, and 6 the opposite way. Now there are 6 isomeric forms of dimethylcyclohexane: cis-1,2; trans-1,2; cis-1,3; trans-1,3; cis-1,4; and trans -1,4. For cis-1,2-dimethylcyclohexane, if the methyl group on 1 is axial then the methyl group on 2 is equatorial. If we switch chairs (chair flip), the axial methyl on 1 becomes equatorial and the equatorial methyl on 2 becomes axial – so we still have one axial and one equatorial! These two chairs are equal in energy. For trans-1,2-dimethylcyclohexane, if the methyl group on 1 is axial, then the methyl group on 2 is also axial. With the chair flip, both become equatorial. The diequatorial chair is lower energy, the diaxial chair is higher energy. For cis-1,3-dimethylcyclohexane, if the methyl on 1 is axial, then the methyl on 3 is axial. The chair flip makes them both equatorial. Again, diequatorial is lower energy, diaxial is higher energy. For trans-1,3-dimethylcyclohexane, if the methyl on 1 is axial, the methyl on 3 is equatorial. The chair flip makes them equatorial and axial. Since in both chairs there is 1 axial and 1 equatorial, they are equal in energy. For cis-1,4-dimethylcyclohexane, if 1 is axial, 4 is equatorial. The chair flip makes them equatorial and axial, and again the two chairs are equal in energy. For trans-1,4-dimethylcyclohexane, if 1 is axial, then 4 is axial. The chair flip makes them both equatorial. The diaxial chair is higher in energy, the diequatorial chair lower in energy. These principles are general – if there are 3, or 4, or more substituents. Draw the two chairs, figure out how many axial and equatorial substituents on each. The more equatorial – the lower the energy. It is a common problem to give you a particular cyclohexane stereoisomer, then ask you to draw the two chair conformations and identify if there is a lower energy conformation (the two conformations may be equal energy!).

chair flip   cis 1,2   chair flip   trans 1,2

chair flip   cis 1,3   chair flip   trans 1,3

chair flip   cis 1,4   chair flip   cis 1,4

Quick Review.

1. CONFORMATIONS: C-C and C-H bonds freely rotate. Changes in molecular shape with C-C rotations are called conformations, analysis of energy changes with rotation is called conformational analysis.

2. ETHANE and NEWMAN PROJECTIONS: Ethane has two types of conformations – eclipsed conformations where the C-H bonds line up, and staggered conformations where they alternate. We use dihedral angles to define these conformations, and Newman projections to draw them. The staggered conformations are low energy, the eclipsed conformations are high energy.

3. BUTANE: ANTI and GAUCHE: Butane has four types of conformations around the central C-C bond. There are two types of eclipsed conformations: one where the two C-C bonds line up (C-C eclipsed) – highest in energy, and two where the C-C and C-H bonds line up (C-H eclipsed) – high in energy. There are two staggered type conformations, two called gauche and one called anti (the lowest in energy with the CH_3 groups opposite to each other.)

4. STRAIN: The increased energy of some conformations relative to others is called strain. The source of strain can be distortions of bond angles (angle strain), bond rotations (torsional strain), or the intrusion of atoms into each other's space (van der Waals strain).

5. CYCLOALKANES: Cyclic compounds show strain. Cyclopropane is the most strained, with high angle strain and high torsional strain. Cyclobutane is also highly strained, cyclopentane mildly strained, and cyclohexane unstrained (as compared to linear alkanes). Strain increases for larger rings.

6. CYCLOHEXANE: Unstrained cyclohexane is the most studied ring system. It folds into a chair conformation with 109.5° angles and staggered conformations. There are two chairs, and high energy conformations called boats. The C-H bonds are either equatorial or axial.

7. SUBSTITUTED CYCLOHEXANE: Methylcyclohexane has two chairs, one with the methyl equatorial, one with the methyl axial. Equatorial substituents are lower in energy. Dimethylcyclohexanes fall into two groups. Some have 1 axial and 1 equatorial substituent. Then the chairs are equal in energy. Others have a chair with 2 axials and a chair with 2 equatorials. Then the di-equatorial chair is lower in energy.

7.   Chirality

1. What is Chirality? Chirality comes from the Greek word cheir (χειρ), which means "hand" – and the simplest meaning of chirality is "handedness". All of you are experts in chirality – you just don't know the lingo! How do I know? Well, we deal with chiral objects every day: do you put your shoes on the correct feet? Do you know how to unscrew a cap from a drink bottle? Then you know chirality! An object is said to be CHIRAL if it is "nonsuperimposable on its mirror image". So think about your hands: the right hand is a mirror image of the left: in the mirror, your right hand looks like a left hand. But the two hands are nonsuperimposable: if you line up the fingers and thumbs correctly, the palms are on opposite sides. If you line up the palms correctly, the fingers are wrong. So this is CHIRALITY: the property of being nonsuperimposable, like hands. In our macroscopic world, a lot of things are chiral – many of the familiar ones have to do with hands, others with spirals: hands, feet, shoes, nuts and bolts, spiral stairs. Some gloves are chiral, schoolroom desks are often chiral (different desks for righties and lefties). Many objects are also ACHIRAL – that means they *can* be superimposed on their mirror image: socks are nearly always achiral, tables, chairs, and on and on. Any object that has an internal mirror plane (plane of symmetry) such that one half of the object reflects onto the other, is always ACHIRAL. So the exterior of the human body, although it has chiral pieces (hands, feet) is itself ACHIRAL.

Organic molecules can be chiral. This is most commonly seen when a tetrahedral carbon has 4 different substituents. In the figure below, the left side shows two mirror image structures, the dashed line between represents the mirror plane. On the right side, the two mirror images are shown again, now the furthest right structure has been rotated 180°. Note that two of the substituents ("Y" and "Z") line up, but that the other two ("W" and "X") do not: "X" is in the back on the left hand structure, "W" is in the back on the right. So we can make any two substituents line up, but the other two are always "wrong" – just like the fingers and palms of your hand! A carbon with 4 different substituents is called a "STEREOGENIC CENTER"(it generates stereochemistry) – they used to be called "chiral centers" but this is no longer approved in modern usage (because - as we will see -it is possible to have these centers but be achiral.)

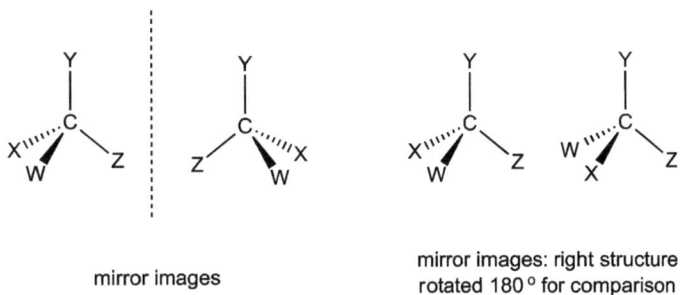

mirror images

mirror images: right structure
rotated 180° for comparison

The two mirror image isomers are STEREOISOMERS – just like the cis and trans stereoisomers we saw above for cycloalkanes. Like all stereoisomers, they have the same molecular formula, and the same connectivity, but differ in the way the atoms are arranged in space. But the difference is even more subtle than the cis-trans isomers – they are mirror images. They are the most similar any isomers can be. The cis-trans isomers usually have different polarities, and different melting points and so on. The nonsuperimposable mirror image isomers have all of the same common physical properties – only special physical properties that involve chirality show any difference. For that reason, these isomers are given a special name: ENANTIOMERS. Enantiomers are those stereoisomers that are nonsuperimposable mirror images. All other stereoisomers are considered DIASTEREOMERS.

It is likely that you will be asked to identify these relationships: are these two compounds identical, enantiomers, or diastereomers?

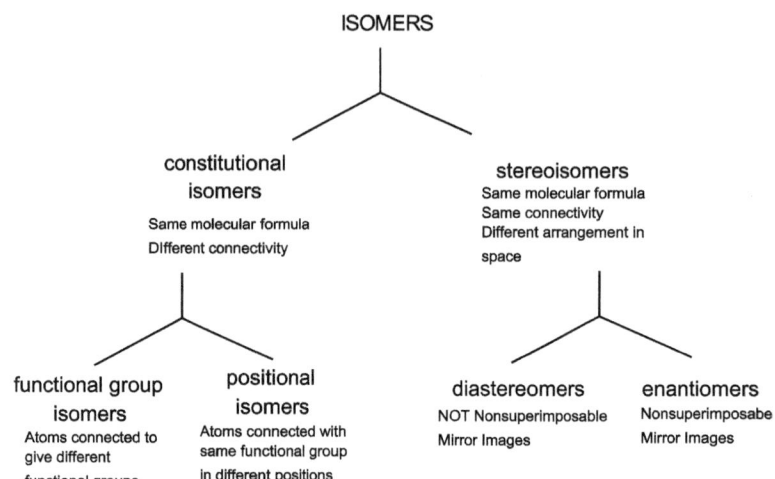

2. Since enantiomers are different compounds, we need a way to distinguish them: like we did with cis and trans for the substituted cycloalkanes. The IUPAC way of distinguishing them is the R,S system (sometimes called Kahn-Ingold-Prelog for its inventors). In the end, a stereogenic center is designated as either R or S: this is called its CONFIGURATION. The R,S system starts with a method for assigning "PRIORITIES" to the groups around a stereogenic center (carbon with 4 different substituents). First, we compare the atomic number (Z) of the first atom of each group. The higher the atomic number, the higher the priority. (The highest priority is "1", the lowest "4".) So if the atoms are Br (Z = 35), Cl (Z = 17), F (Z = 9), and H (Z = 1), then the priorities are Br-1, Cl-2, F-3, and H-4. What if two atoms are the same (most commonly, both carbons)? Then we compare the atoms attached to those two atoms, and any difference in atomic number gives the higher priority to the higher atomic number. So if we have a methyl (-CH_3) and ethyl (-CH_2CH_3), both have C (Z=6) attached to the stereogenic center. The next attached atoms are H,H,H for methyl, C,H,H for ethyl : the C on the ethyl has a higher atomic number, so ethyl is higher priority. For linear alkyl groups, a longer chain is always higher priority. Branching increases the priority: if we compare propyl and isopropyl, both have C attached to the stereogenic center, and the attached atoms are C,H,H for propyl but C,C,H for isopropyl: the second C makes isopropyl the higher priority. If there is no difference between two groups, they are identical: if there are two identical groups on the carbon, it is not a stereogenic center! There are a few other more esoteric rules, but these cover most situations.

Once we have assigned all of the priorities, we can assign the configuration. We must rotate the stereogenic center so that the lowest priority (4) is in the back. Then we trace an arc from priority 1 to 2 to 3. If this arc goes CLOCKWISE ↻, the configuration is R. If the arc goes COUNTER-CLOCKWISE ↺, the configuration is S. The hardest part of this is the rotation of the stereogenic center to the proper orientation. There are two methods that are found to work well. One is to rotate the molecule around one of the four bonds – this bond "axis" is unchanged, but the other three move around the axis. The second way is to recognize that if we "switch" two of the groups, we convert the molecule into its enantiomer: R into S. If we do a second switch, S turns back into R! If the result of this "double switch" puts the lowest priority in back, then we have reoriented the molecule! Either method works to give the name (if done correctly) – it all depends on how YOUR brain works! The configuration label is given before the name: (R)-3-methylhexane.

You need to know how to name stereogenic centers as R or S!

There are some other naming conventions for enantiomers in use: an older system (the D,L system) is still used for sugars and amino acids, and a special property of chiral compounds means one enantiomer can be called (+) and the other (-): details to follow!

Rotation around axis method:

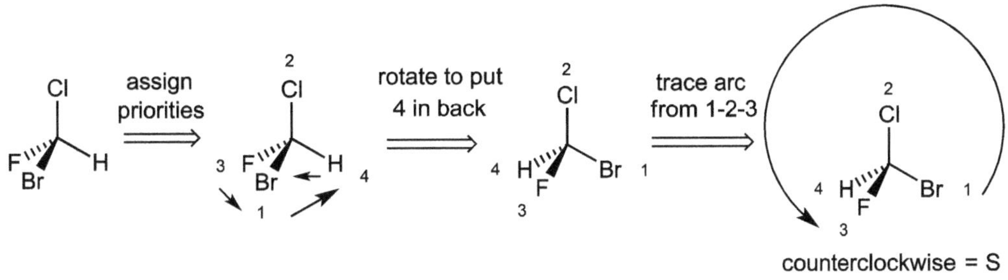

assign priorities ⟹ rotate to put 4 in back ⟹ trace arc from 1-2-3 ⟹

counterclockwise = S

Double Switch method:

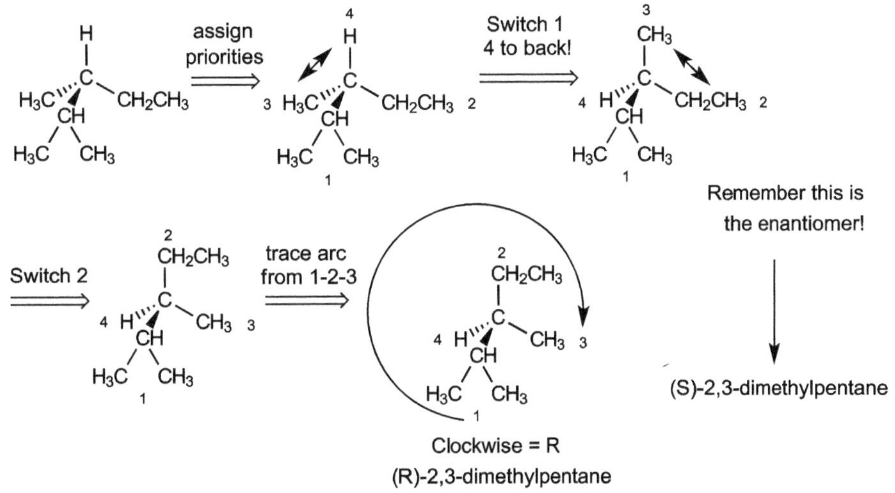

assign priorities ⟹ Switch 1 4 to back! ⟹

Remember this is the enantiomer!

Switch 2 ⟹ trace arc from 1-2-3 ⟹

(S)-2,3-dimethylpentane

Clockwise = R
(R)-2,3-dimethylpentane

3. If there is a single stereogenic center in a molecule, there are two enantiomers – R and S. If there is a second stereogenic center, it can also be R or S: so it doubles the number of stereoisomers: R,R or R,S; S,R or S, S. A third stereogenic center doubles the number again – 8 possibilities – RRR, RRS and so on. If there are N stereogenic centers, there can be as many as 2^N stereoisomers – CAN BE: sometimes there are less! When are there less? If a molecule has multiple stereogenic centers AND the possibility of an internal mirror plane – when the internal mirror plane occurs, two of the possible stereoisomers become achiral and collapse into a single compound: this is called a MESO compound. Meso compounds only occur if the molecule can have an internal mirror plane – that is why for 2 stereogenic centers there may be 4 stereoisomers or 3 (with the meso). This is just like the human body – the hands and feet are stereogenic centers, but they reflect onto each other so overall the body is achiral.

The multiple stereoisomers of a molecule with multiple stereogenic centers occur in pairs of enantiomers: R,R is the enantiomer of S,S; R,S is the enantiomer of S,R (unless these two become the meso!). What is the relationship between R,R and R,S? They are not enantiomers – and any stereoisomers that are not enantiomers are DIASTEREOMERS. Remember that "enantiomer" and "diastereomer" are names of relationships – like "brother", "sister" or "cousin" – so just as an individual is one person's sister but another person's cousin, so in a "family" of stereoisomers, a molecule is the enantiomer of one stereoisomer, but a diastereomer of others.

4. Optical Rotation: As mentioned above, enantiomers are the most similar of all isomers – mirror images. Most of their properties are the same – but they are different in a chiral environment. The most common way this can be seen is through the interaction of the enantiomers with polarized light: this is called POLARIMETRY. Light is made of waves. In a beam of normal (unpolarized) light, the light waves vibrate in all directions. If light is passed through a polarizing filter, only one direction (plane) of vibration makes it through the filter: this is plane polarized light. If plane polarized light passes through a chiral substance, the plane of polarization is ROTATED. The rotation can be to the right (clockwise, designated (+)) or to the left (counterclockwise, designated (-)). If a given compound rotates the plane to the right (+), then its enantiomer will rotate the plane an equal amount to the left (-): so one enantiomer is (+) and one is (-). When the absolute configuration (R or S) is unknown, the rotation can be measured, and the enantiomers designated as (+) and (-) until the absolute configuration is known!

Just like with UV-Vis absorbance spectroscopy, the amount of rotation depends on the wavelength of light, the concentration of the sample, and the pathlength of the sample. The observed rotation is described using the symbol α. The SPECIFIC ROTATION is described by putting the α in brackets: $[\alpha]$. The definition of SPECIFIC ROTATION is based on the concentration "c" in g/mL and the pathlength "l" in decimeters (dm). The wavelength and temperature are also specified – the usual wavelength is from a sodium vapor lamp – the "D" line is an orange color at 589 nm (you have seen it in street lamps) – and is shown by a superscript "D". The temperature is often given as a subscript – 20 °C or 25 °C are most common. Specific Rotation is the polarimetry version of the extinction coefficient in UV-Vis.

$$[\alpha]^D_{20} = \frac{\alpha}{c * l}$$

So for a simple example: if a sample of 0.5 g/mL of a substance has a measured α of +10 °, using a cell with a pathlength of 1 dm (this is most common), then the specific rotation $[\alpha]$ is +10/(0.5) or +20 °. You will probably be expected to do this kind of calculation.

Polarimetry is the most common measurement done on chiral substances, but there are others: One can measure the polarimetry as a function of wavelength – this is called OPTICAL ROTATORY DISPERSION – or use a more complex type of polarized light (circularly polarized) in a technique called CIRCULAR DICHROISM or CD: these techniques give more information on the enantiomers, but require more expensive instrumentation. To experimentally determine the absolute configuration of a specific stereoisomer is even more difficult, and usually requires certain techniques of X-Ray crystallography.

The old D,L system of nomenclature was based on polarimetry. The simplest sugar is glyceraldehyde, which has a single sterogenic center, so there are two enantiomeric forms. Only one glyceraldehyde enantiomer exists in nature. All other naturally occurring sugars can be shortened, one carbon at a time, until they become the naturally occuring glyceraldehyde. The natural glyceraldehyde rotates plane polarized light to the right, so the old Latin derived term was DEXTROROTATORY. Therefore it was labelled "D". The glyceraldehyde that does not occur naturally rotates left – LEVOROTATORY – so it was labelled "L". Since the other natural sugars are the same as glyceraldehyde at that particular stereogenic center, they are all considered "D" sugars (i.e. D-glucose). Their enantiomers are considered "L" (L-glucose does not occur in nature). NOTE WELL – the "D" designation means "same configuration as D-glyceraldehyde" – some of the "D-sugars" may themselves be levorotatory. If you arrange the natural amino acids to correspond to glyceraldehyde, they resemble the "L" glyceraldehyde, so they are considered L amino acids. With other compounds it is not always easy to make them correspond to glyceraldehyde, so the R,S system was devised and mostly replaced D,L – the old system only survives in biochemistry for sugars and amino acids.

5. Enantiomers are the most similar of isomers. They have all the same physical properties: MP, BP, density, solubility, etc. in an achiral environment. They only differ in an environment that reflects chirality: solubility in a chiral solvent, or binding energy to another chiral substance, or rotation of plane polarized light. Without the intervention of another chiral object, it is impossible to tell enantiomers apart. (Remember feet in shoes vs feet in socks!) Diastereomers, however, have DIFFERENT physical properties: MP, BP, etc. The differences may be small (or large), but they always exist. So it is IMPOSSIBLE to separate ENANTIOMERS by distillation, but it is POSSIBLE (although perhaps difficult) to separate DIASTEREOMERS by distillation!

There is one mixture of enantiomers that is special. That is the mixture which is exactly 50:50 – half R and half S, half (+) and half (-). This mixture is called a RACEMIC MIXTURE or RACEMATE, and is sometimes given the symbol (±). This mixture is special because it acts as if it were ACHIRAL – it does not rotate the plane of polarized light – so it is said to be OPTICALLY INACTIVE. Why? Well, every time a (+) molecule rotates the plane to the right, there is a (-) molecule that rotates it an equal amount left, and they cancel! So unless you know it is a mixture, you think it is achiral! The mystery writer Dorothy Sayers wrote a book ("The Documents in the Case") where the dramatic revelation that a death was murder came when the poison was tested by polarimetry: if it was an accidental death (mushroom poisoning), the poison would be chiral; if deliberate murder (addition of synthetic poison), then the poison would be racemic and achiral. What happened? Read the book....

How do we evaluate other mixtures of enantiomers? The raw measurement is the % OPTICAL PURITY: we measure the optical rotation α, determine the specific rotation $[\alpha]$, and compare to the specific rotation known for the pure compound.

$$\% \ Optical \ Purity \ = \ \frac{[\alpha](measured)}{[\alpha] \ (pure)} \ x \ 100\%$$

So if the pure enantiomer has $[\alpha]$ of -40 °, and we measure $[\alpha]$ of -30 °, then the % Optical Purity is: -30/-40 x 100% = 75%. Note that the measured and pure $[\alpha]$ should have the same sign. (Also note that racemic mixtures have an optical purity of 0!) The other way of describing the purity of a mixture of enantiomers is by % ENANTIOMERIC EXCESS (sometimes just called enantiomeric excess, or ee). The enantiomeric excess is formally described by comparing the molarities of the two enantiomers: the enantiomer present in greater amount is 1, in lesser amount is 2. Because the units cancel, any concentration measure will work: g, mol, g/mL, mol/L, etc.

$$\% \ enantiomeric \ excess = \frac{[enantiomer \ 1] - [enantiomer \ 2]}{[enantiomer \ 1] + [enantiomer \ 2]} \ x \ 100\%$$

It turns out that % Optical Purity and ee are IDENTICAL! An ee of 75% will give a % optical purity of 75%! (and vice versa) One key concept to realize is that for both measurements, the IMPURITY IS THE RACEMIC MIXTURE. If the % optical purity is 75% (-), the other 25% is racemic: 12.5% (+) and 12.5% (-). So overall, the composition is 87.5% (-) and 12.5% (+). What is the ee, then? (87.5 – 12.5) / (87.5 + 12.5) = 75/100 = 75% ee! In both measures THE IMPURITY IS THE RACEMIC MIXTURE!

It is very likely that you will be asked to calculate ee or % Optical Purity.

The separation of a mixture of enantiomers (racemic or not) is called a RESOLUTION. The old fashioned, traditional way is to take the mixture, (R) and (S), and do a reversible reaction with the pure enantiomer of another substance: if this is all (R) then you have (RR) and (RS) products – which are diasteromers! Diastereomers can be separated by distillation or recrystallization, so these are separated. Reversing the reaction gives the pure (R) and (S) enantiomers of the original mixture. The simplest way to do this is if the mixture of isomers is either acid or base, and the pure enantiomer is the opposite - base or acid – so the diastereomeric mixture is a salt. This salt formation is easily reversible, since no actual covalent bonds are formed. More modern methods of resolution include chiral chromatography columns, but the old fashioned way is still best when a large quantity of material needs to be resolved.

6. Chemical reactions of stereoisomers. Chemical reactions of chiral compounds have some special properties as well. What if a single stereoisomer (one of two enantiomers) reacts? There are three possible outcomes: the stereochemistry of the product could be the same (RETENTION) or it could be the opposite (INVERSION) or a mixture could be produced (RACEMIZATION). If a reaction proceeds through an achiral intermediate, racemization will always result. We have to be a little bit careful about retention and inversion: an "R" starting material can give "R" product but the result could be inversion. How? Because the R and S designations are arbitrary: inversion could occur but if at the same time the arbitrary order of priorities are reversed, then the designation is still "R"! We have to compare the actual starting and ending configurations to see if it is Retention or Inversion. There are some other terms that are used for describing stereochemistry of reactions. A reaction is STEREOSELECTIVE if it produces more of one stereoisomer than of another. This description includes *cis/trans* diastereomers – a reaction that produces more of the *trans* than the *cis* is stereoselective. The stereoselectivity can be complete (100% of one isomer) or partial (1-99%). Many drug substances are chiral, and methods that produce only (or mostly) the desired isomer are preferred, so stereoselective reactions are the focus of a lot of research. A reaction is STEREOSPECIFIC if one stereochemical form of the starting material gives one stereochemical form of the product, while a different stereochemical form of the starting material gives a different stereochemical form of the product. The classic example is the bromination of 2-butene. Bromination of cis-2-butene gives a racemic mixture of 2,3-dibromobutanes (2R,3R and 2S,3S), while bromination of trans-2-butene gives the meso 2,3-dibromobutane (2R,3S). Note that STEREOSELECTIVE only compares stereoisomeric products from a single starting material, while STEREOSPECIFIC compares two reactions – 2 different starting materials, 2 different products.

7. The disubstituted cyclohexanes exhibit a wide variety of stereochemical behavior, and add the concept of dynamics (motion, changes) from their changing conformations. Let us use the dimethyl cyclohexanes to show these effects. 1,2-dimethylcyclohexane can be *cis* or *trans*. The two substituted carbons of the ring are both stereogenic centers, with 3 different substituents: H, -CH$_3$, ring toward the other –CH$_3$, ring away from the other – CH$_3$. It is useful in showing the cis-trans relationships to use a special drawing of the ring – a Haworth projection – that envisions the ring as planar; however, this drawing is also deceptive, since the ring is not actually planar – it exists as two interconverting chair forms. The Haworth projection makes it appear that the *cis* 1,2-dimethylcyclohexane has an internal mirror plane and is achiral. When we make the chair conformations however, one methyl group is axial and one is equatorial: it does not have an internal mirror plane! We can make the mirror image, and we find that the mirror images are nonsuperimposable so they are chiral – enantiomers. However, if we cause the chair to undergo a chair flip to the other chair, this chair is the enantiomeric chair! So while the *cis*-1,2 is not achiral, it exists in equilibrium with its enantiomer (the other chair) as a racemic mixture. Whew! That's complicated! The trans-1,2-dimethylcyclohexane is easier: there are two enantiomers, and each

exists as two chairs: a di-axial chair and a di-equatorial chair. The details of all of the isomers are summarized in the table below; this is a little exotic but if your professor covers it and you want the A, you need to know it!

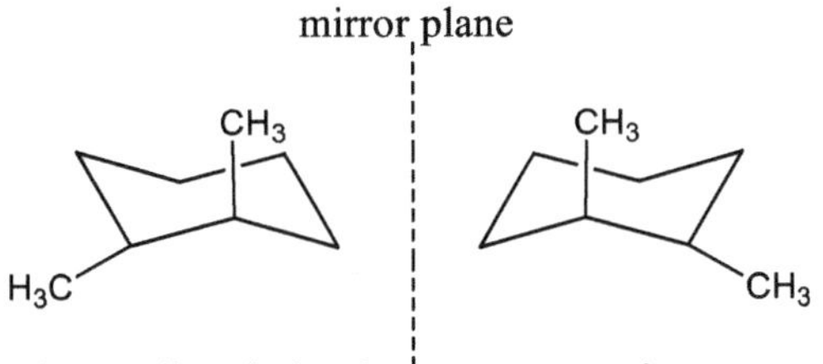

mirror plane

cis-1,2-dimethylcyclopropane enantiomers

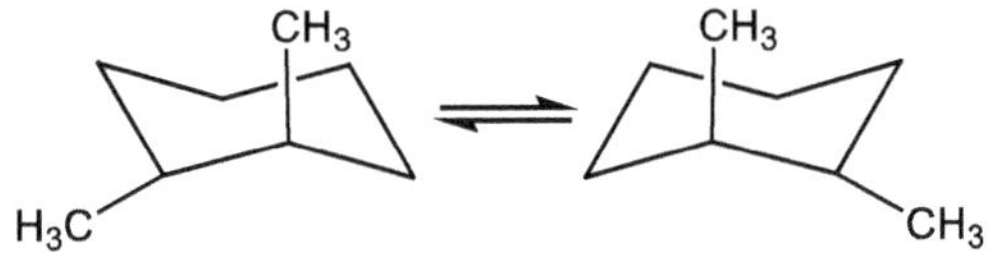

enantiomers interconvert: racemic equilibrium mixture

Dimethyl-cyclohexane	Haworth Projection	Expected to be chiral?	Chair Conformation	Is chair chiral?	Details
Cis-1,2 1R, 2S	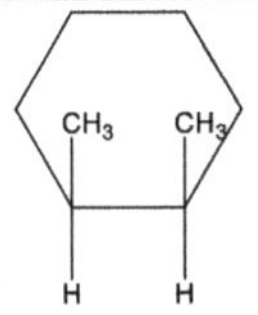	No	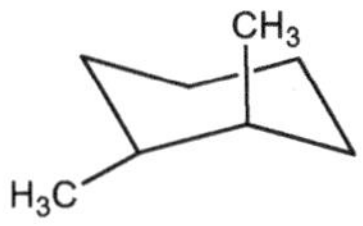	Yes	Chair flip interconverts enantiomeric chairs: equilibrium mixture is racemic
Trans-1,2 1R,2R & 1S,2S	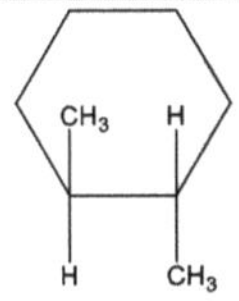	Yes	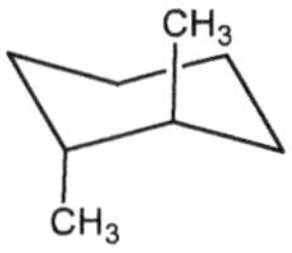	Yes	Two enantiomers, each with a diaxial and diequatorial chair
Cis-1,3 1R, 3S	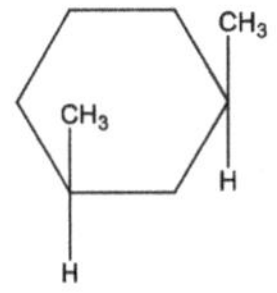	No	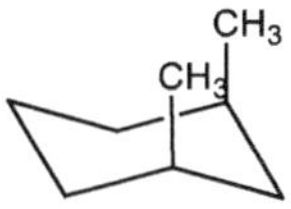	No	Internal mirror plane, chair is meso compound
Trans-1,3 1R,3R & 1S,3S	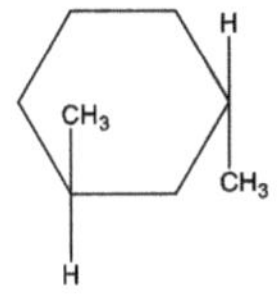	Yes	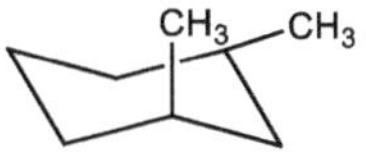	Yes	Each enantiomeric chair is in equilibrium with an identical, superimposable chair
Cis-1,4 achiral	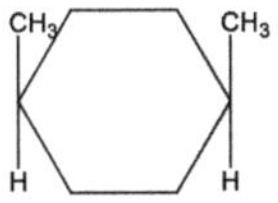	No		No	Internal mirror plane, chair is achiral
Trans-1,4 achiral	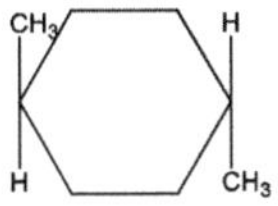	No	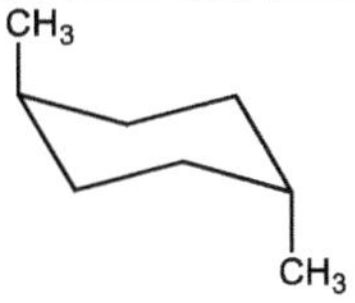	No	Internal mirror plane, chair is achiral

Quick Review

1. CHIRALITY: Chirality is handedness: objects are chiral if they are nonsuperimposable on the mirror image. The two mirror image isomers are called ENANTIOMERS. Stereoisomers that are not enantiomers are considered DIASTEREOMERS. Chirality in organic compounds usually occurs when carbons have 4 different groups attached.

2. R/S NOMENCLATURE: Enantiomers are named by adding the R,S system to the regular nomenclature. The 4 groups around a stereogenic center (carbon) are assigned priorities, the lowest priority is oriented to the back, and a curve is traced from highest priority to second to third: if this curve goes clockwise it is R, counterclockwise it is S.

3. DIASTEREOMERS: If there are multiple stereogenic centers, each is given an R or S designation.

4. POLARIMETRY: Chiral substances have OPTICAL ACTIVITY – they rotate the plane of plane polarized light. The measured rotation is given the symbol α and can be + (clockwise) or – (counterclockwise). The rotation corrected to a standard set of conditions (1 dm pathlength, 1 g/mL concentration) is the specific rotation $[\alpha]$ – the temperature and wavelength of light are usually given as well.

5. RACEMIC MIXTURES Mixtures of enantiomers are defined by two measures – the % optical purity and the enantiomeric excess. These measures are the SAME – a mixture of 20% optical purity will have 20% enantiomeric excess – but are defined differently. Optical purity is based on measurement of specific rotation, enantiomeric excess is defined on known composition. The 0% optical purity/ 0% enantiomeric excess mixture is special: the RACEMIC MIXTURE. It appears to be achiral.

6. REACTION STEREOCHEMISTRY: Chemical reactions of chiral molecules can have three end-member outcomes: RETENTION when the configuration of the stereogenic center stays the same, INVERSION when the configuration changes to the opposite configuration, RACEMIZATION when the stereogenic center is formed equally in both configurations. Reactions are called STEREOSELECTIVE if one stereochemical form is made preferentially, STEREOSPECIFIC if different stereochemical forms of starting material generate different stereochemical forms of product.

7. DISUBSTITUTED CYCLOHEXANES: The stereochemistry of disubstituted cyclohexanes shows a variety of unusual relationships relating conformations and chirality

III. Functional Groups I
8.   Haloalkanes

1. Haloalkanes or alkyl halides are compounds where an alkane has halogen atoms (fluorine, chlorine, bromine, or iodine) replacing one or more hydrogens of an alkane. The functional group is a halogen atom - F, Cl, Br or I. Because of the similarities in properties and chemistry, these are usually treated together – although compounds with F are a little different. The halogens as a group are sometimes represented by "X" – so R-X represents a general alkyl group (R) connected to a general halogen atom (X).

The carbon-halogen bonds are polar, so haloalkanes are slightly polar. The polarity is minimal, so haloalkanes generally mix with nonpolar compounds like alkanes, not highly polar ones like water. The atomic weights of chlorine (35.5), bromine (80) and iodine (127) also make the molecular weights of the haloalkanes fairly high. (Chloromethane, chloroethane, and bromomethane are gases at room temperature, most other haloalkanes are liquids at room temperature.)   Between the polarity and the molecular weight, most haloalkanes have moderately high melting points and boiling points. The one exception is fluorine (mass of 19), which is lighter. Fluoroalkanes can have BP and MP not too different from the hydrocarbon "parents". Haloalkanes do not hydrogen bond well, and with only a little polarity, they are insoluble in water. One notable property of haloalkanes is that many have fairly high densities from the high MW. (Tribromomethane is used by geologists to "float" lighter minerals –ROCKS! - away from heavier ones!) While haloalkanes are effectively nonpolar, they tend to be "polarizable". The electrons on larger atoms with many electrons are easier to attract or repel with nearby charges – causing dipoles to be "induced" – this is polarizable. Cl, Br and I all have this property. This makes the haloalkanes good solvents for nonpolar to moderately polar compounds. In particular, dichloromethane (methylene chloride), trichloromethane (chloroform) and carbon tetrachloride have been used in this way. The manufacture of carbon tetrachloride, however, was banned as an ozone depleting chemical in the 1990's, and the cost of this excellent solvent is now astronomical.

HALOALKANE PHYSICAL PROPERTIES:
Polarity: slightly POLAR to NONPOLAR
MP/BP: moderately low
H_2O sol.: No
Density: High, >1 g/mL

2. The IUPAC nomenclature of haloalkanes is easily explained. Halogen atoms are treated as substituents on the alkane chain. Fluorine is fluoro, chlorine is chloro, bromine is bromo, iodine is iodo. We treat these four new substituents JUST LIKE THE ALKYL GROUPS WE LEARNED IN ALKANE NOMENCLATURE! The halogen groups are mixed in with the alkyl groups and alphabetized with them. They all count the same for figuring out numbering as well. Here are a few examples.

2-chloro-2-methylbutane   2-bromo-3-chloro-3,4-dimethylhexane

The older system named haloalkanes as alkyl halides (as if they were salts (like sodium chloride) with an alkyl group cation and halide anion). This is fine when the alkyl groups are simple ones like methyl, ethyl, etc. But it fails when the alkyl groups are complex. The old system is still used for simple compounds like ethyl chloride or methyl iodide. Your organic professor will expect *you* to use the IUPAC nomenclature for haloalkanes, but you may run into the old system here and there.

3. As we describe these other types of functional groups, we will find that they have richer chemistries than the alkanes. There are chemical reactions where the functional group is either reactant or product. We will learn these reactions in THREE WAYS. We will learn the reactions that prepare ("synthesize") the functional group: the functional group appears on a product. We will learn the reactions of the functional group: the functional group appears on a reactant. Finally, we will learn both types of reactions in terms of their MECHANISMS: the detailed, step-by-step description of the chemical transformation. Often, several functional groups will take part in reactions of the same type that follow the same mechanism – so if we learn the mechanism, we can collect together the information about different functional groups.

How are haloalkanes prepared? We have already seen the halogenation of alkanes. If we mix an alkane and either chlorine gas or bromine (liquid at room temperature, but lots of vapor above it) and shine ultraviolet light on the mixture, after a while we will form haloalkane (R-X) and hydrogen halide (H-X). We will deal with the details and reasons when we take up radical chemistry shortly, but for now it is useful to know that chlorine (Cl_2) is relatively indiscriminate, replacing hydrogens all over any alkane structure. Bromine, however, is selective. It replaces tertiary hydrogens more effectively than secondary or primary, and secondary more effectively than primary. So propane brominates nearly exclusively at the 2 position to make 2-bromopropane (~98%), but chlorinates nearly equally to form 1-chloroporpane and 2-chloroporpane. 2-Methylbutane brominates nearly exclusively to make 2-bromo-2-methylbutane (the tertiary bromide).

$$H_3C\text{---}CH_2\text{---}CH_3 \ + \ Cl_2 \ \xrightarrow{h\nu} \ H_3C\text{---}CH_2\text{---}CH_2Cl \ + \ H_3C\text{---}\overset{\displaystyle Cl}{\overset{|}{CH}}\text{---}CH_3$$

$$H_3C\text{---}CH_2\text{---}CH_3 \ + \ Br_2 \ \xrightarrow{h\nu} \ H_3C\text{---}\overset{\displaystyle Br}{\overset{|}{CH}}\text{---}CH_3 \ (98\%)$$

We can also make haloalkanes from alcohols. Certain reagents will convert the alcohol –OH to –X. The hydrogen halides (HCl, HBr, HI) will do this, but there are certain drawbacks to these reactions. Thionyl chloride ($SOCl_2$) will convert an alcohol to a chloride, and phosphorus tribromide (PBr_3) will convert an alcohol to a bromide. These will be covered again as reactions of alcohols – but the thionyl chloride reaction is shown.

$$H_3C\text{---}CH_2\text{---}OH \ \xrightarrow{SOCl_2} \ H_3C\text{---}CH_2\text{---}Cl$$

Finally, we can make haloalkanes from alkenes, by addition of the hydrogen halides (HX). In an ADDITION reaction, a molecule adds its two components (here H^+ and X^-) to the two carbons of the double bond. In some cases, there are two possible products, where X^- adds to the two different carbons of the double bond. Often, only one of these products is formed, following what is known as the "Markovnikov Rule". These details will be covered later.

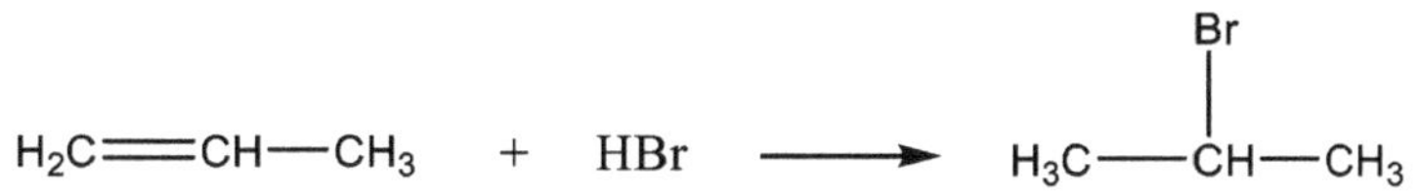

$$H_2C = CH - CH_3 \quad + \quad HBr \quad \longrightarrow \quad H_3C - \underset{\underset{Br}{|}}{CH} - CH_3$$

4. What reactions are common for the haloalkanes? Two of the central reactions of first term organic chemistry are reactions of haloalkanes. These two reactions are NUCLEOPHILIC SUBSTITUTION (S_N) and ELIMINATION (E). In Nucleophilic Substitution, a reagent with a lone pair of electrons (the nucleophile – one example is H_2O) replaces the halogen; the halogen leaves the haloalkane as the halide anion (therefore the halogen is called a "leaving group"). The NUCLEOPHILE SUBSTITUTES for the halogen. In Elimination, the halide group leaves as halide ion, and a base removes a proton (H^+) from an adjacent carbon. The electrons that made a bond to the proton shift over and form a double bond between the two carbons. The small molecule HX has been ELIMINATED from the reactant. Both Nucleophilic Substitution and Elimination have a lot of variation and detail – but we will cover those later. You should recognize these two reactions – we will show later how to predict which one will happen.

SN: $\quad H_3C - Br \quad + \quad H_2O \quad \longrightarrow \quad H_3C - OH \quad + \quad Br^- \quad + \quad H^+$

E: $\quad H_3C - CH_2Cl \quad + \quad OH^- \quad \longrightarrow \quad H_2C = CH_2 \quad + \quad Cl^- \quad + \quad H_2O$

Another minor reaction is the generation of organometallics: magnesium and lithium metals will replace halogens to make alkyl magnesium and alkyl lithium compounds.

$\quad H_3C - Br \quad + \quad Mg \quad \longrightarrow \quad CH_3MgBr \qquad\qquad H_3C - Br \quad + \quad Li \quad \longrightarrow \quad CH_3Li \quad + \quad LiBr$

5. What are some of the uses of haloalkanes? Starting materials for the reactions shown above, solvents, anesthetics, refrigerants… From the reactions we described, they are starting materials for a variety of reactions to make other compounds. The smaller haloalkanes – dichloromethane, trichloromethane (chloroform), 1,2-dichloroethane, and so on – are often used as solvents. You may occasionally have heard of chloroform as an anesthetic – in movies, people apply a chloroformed cloth to someone's mouth to knock them out (it actually takes a while in real life). Chloroform used to be used as an anesthetic, but it can cause liver damage, so halothane (2-bromo-2-chloro-1,1,1-trifluoroethane) or certain fluorinated ethers are now used when an inhaled anesthetic is required. Another use of haloalkanes is as refrigerants. The chlorofluorocarbons (CFC's or Freons) were discovered in the 1930's, and found to be both relatively unreactive (inert) and possessed of a high heat capacity, desirable in the working fluid in refrigerators and air conditioners. Unfortunately, the Freons were later found to be ozone depleting chemicals: they drift to the stratosphere, where the chlorine atoms are knocked off by ultraviolet radiation. The released chlorine atoms then catalyze the destruction of ozone. The manufacture of these chemicals was banned in the 1990's, and they were replaced by hydrochlorofluorocarbons (HCFC's), which are degraded in the atomosphere before they can reach the ozone layer. If you have a very old air-conditioner, it may use the old Freons. If you need to recharge it, it is very expensive: manufacture was made illegal, but existing supplies could be used, but as they grow smaller the law of supply and demand kicks in…. but that is a science even more dismal than this one.

Only 5 points for this topic!

Quick Review

1. HALOALKANES: Haloalkanes are compounds where a halogen has replaced a hydrogen on an alkane. They are nonpolar, but have moderate BP and MP due to the high mass of most halogens. They are insoluble in water and tend to have higher densities.

2. NOMENCLATURE: IUPAC names haloalkanes as substituted alkanes. Each halogen has a name as a substituent (fluoro, chloro, bromo, iodo) and the rules for naming substituted alkanes apply.

3. PREPARATION: Haloalkanes can be prepared from alkanes: alkane + halogen (Cl_2 or Br_2) with light or heat. They can also be prepared from alcohols using special reagents (thionyl chloride $SOCl_2$ or phosphorus tribromide PBr_3). They can be prepared by addition of hydrogen halides to alkenes.

4. REACTIONS: Haloalkanes are the starting materials for many reactions. Notable for us are Nucleophilic Substitution and Elimination.

5. USES: Haloalkanes have a variety of uses, including as refrigerants and anesthetics.

9.   Alkenes

1. The alkenes are compounds with a carbon-carbon double bond. This is their functional group. The alkenes are also hydrocarbons – only carbon and hydrogen. They are UNSATURATED hydrocarbons, however – there are fewer hydrogens than the alkanes. For a double bond, we lose two hydrogens. So the general formula of an alkene is C_nH_{2n}. Since they are hydrocarbons, all the bonds are nonpolar, and the alkenes are therefore also nonpolar. This means they have low MP and BP – pretty much the same as the alkanes. Like the alkanes they are also insoluble in water, and have low densities.

ALKENE PHYSICAL PROPERTIES

Nonpolar
Low MP/BP
Insoluble in water
Low densities

2. The nomenclature of the alkenes requires us to have a new suffix: -ene replaces –ane to indicate the presence of the double bond. We also need to locate the double bond. The simplest alkenes (ethene and propene) don't require a number to locate the double bond – there is only one place it can go! But for alkenes longer than propene (butene and longer), the double bond could be in different positions along the chain – at the end, or at different places in the middle of the chain. So here are the rules for naming alkenes: RULE 1. Locate the longest chain that includes both carbons of the double bond. This is the parent chain. The name of the alkene will be based on the alkane of this length with the suffix changed from –ane to –ene. So a two carbon chain with a double bond is *ethene*. RULE 2. Number the chain so the double bond has the lower set of numbers. There are 2 carbons for the double bond, so there are 2 numbers associated with the double bond. In some cases, the double bond will be right in the center of the chain, so it will not make any difference which way we number. RULE 3. Use the *lower* of the two numbers to indicate the location of the double bond. So if there is a 4 carbon chain with the double bond at one end, the double bond is between carbons 1 and 2. We use the lower number – 1 – to indicate the location: this is 1-butene. If there is a 4 carbon chain with the double bond in the middle, it is between carbons 2 and 3, and we use "2": this is 2-butene. RULE 4. If there are substituents, list them by name and number before the rootname, the same as the alkanes. For those alkenes where the double bond is in the center, if there are substituents, number to give the lower number at the first point of difference (just like the alkanes). Use this numbering for RULE 3. These rules allow us to name the alkenes. A few small alkenes have old, trivial names: ethene is ethylene, propene is propylene. As always, expect questions on nomenclature.

H_2C══CH──CH_2──CH_3

1-butene

H_3C──$\overset{\displaystyle CH_3}{\underset{\displaystyle |}{C}}$══$CH$──$CH_2$──$CH_3$

2-methyl-2-pentene

3. As we saw in the description of bonding using orbitals, the hybridization model describes the double bond as a sigma bond made between sp^2 hybrid orbitals on each carbon, and a pi bond made between two p orbitals. This makes rotation of the double bond a very high energy event – we have to break the pi bond to rotate. As with the cycloalkanes, limited rotation of the double bond allows for stereoisomers - cis-trans isomerism. The requirement

for cis-trans type isomers of alkenes is that both carbons of the double bond must have two different substituents attached. So in 1-butene, where carbon 1 has two hydrogens and carbon 2 has one hydrogen and one –CH$_2$CH$_3$ group, there are no cis-trans isomers: carbon 1 has two identical substituents. But for 2-butene, carbon 2 has one H and one –CH$_3$, and carbon 3 has one H and –CH$_3$. Both have two different substituents, so there are cis-trans isomers. For alkenes, it is called cis if two identical groups are on the same side of the double bond, trans if two identical groups are on the opposite sides of the double bond. Most commonly, the identical groups are hydrogens, but they can be other types of groups. So we can have *cis*-2-butene where the hydrogens are on the same side of the double bond and *trans*-2-butene where the hydrogens are on opposite sides of the double bond. The cis-trans nomenclature is often used for these alkenes, but there are similar isomers where the cis-trans nomenclature can't be used: there are different groups on each of the double bond carbons, but not two that are the same on both carbons. So if we have an ethene (2 carbon chain with a double bond) where one carbon has a hydrogen and a fluorine and the other carbon has a chlorine and a bromine (1-bromo-1-chloro-2-fluoroethene), there are two isomers: one where the fluorine and bromine are on the same side, one where the fluorine and bromine are on opposite sides. The IUPAC has a way to name these isomers – the E/Z system. For each carbon, we assign "priorities" to the two groups that are attached – using the same system we used for the R/S names for enantiomers. After we have identified the higher priority group on each carbon, we can name the two isomers. If the two higher priority groups are on the same side of the double bond, it is the Z isomer (Z for the German word *zusammen* = "together"); if the two higher priority groups are on opposite sides, it is the E isomer (E for *entgegen* = "opposite"). So in the case of 1-bromo-1-chloro-2-fluoroethene, the bromine is the higher priority on carbon 1, and fluorine is the higher priority on carbon 2. When the bromine and fluorine are on the same side, it is (Z)-1-bromo-1-chloro-2-fluoroethene. When the bromine and fluorine are on opposite sides, it is (E)-1-bromo-1-chloro-2-fluoroethene. In most cases, the isomer that is *cis* in the old nomenclature is Z, the isomer that is *trans* is E. So in the common case where the two identical groups for cis/trans are hydrogens, the hydrogens are always the lower priority group, and *cis* is always Z and *trans* is always E. However, it is possible that the identical groups are higher priority on one C and lower priority on the other – as in 1-bromo-1,2-dichloro -2-fluoroethene. The chlorine is lower priority than bromine, but higher priority than fluorine. So two chlorines on the same side is *cis*, but E; while chlorines on opposite sides is *trans*, but Z. You will need to be able to apply the E/Z system – expect questions on it.

(Z)-2-butene
cis-2-butene

(E)-2-butene
trans-2-butene

(Z)-1-bromo-1-chloro-2-fluoroethene

(E)-1-bromo-1,2-dichloro-2-fluoroethene
cis-1-bromo-1,2-dichloro-2-fluoroethene

4. Alkenes can be included in cyclic structures. These are then called CYCLOALKENES. The names of unsubstituted cycloalkenes are made just by adding the prefix "cyclo". So a 3 membered ring with a double bond is cyclopropene, a 4 membered ring with a double bond is cyclobutene, the 5 membered is cyclopentene, the 6 membered is cyclohexene, and so on. Since there are no ends, we can begin numbering anywhere. So we always make one of the double bond carbons "1", and the other double bond carbon "2". Since they are always carbons

1 and 2, we do not need a number – all cycloalkenes are 1-cycloalkenes. It isn't exactly WRONG to add the 1, but if you call it 1-cyclohexene it sounds a bit foolish. If the cycloalkene has substituents, the double bond carbons are still 1 and 2, but the numbering is chosen to give the substituents the lower number at the first point of difference. So the structure shown is 1,5 dimethylcyclopentene – the double bond has to be 1 and 2, so the double bond C with the methyl is 1. The other methyl then falls at C5.

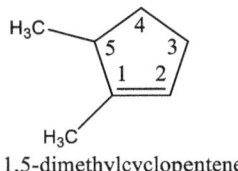
1,5-dimethylcyclopentene

Like cycloalkanes, cycloalkenes are often strained. The double bond carbons have bond angles of 120°, so the internal angles of cyclopropene (60°), cyclobutene (88°), and cyclopentene (104°) cause angle strain. Cyclohexene again has no angle strain.

Cycloalkenes of reasonable size are always cis. In order to incorporate a trans double bond requires a ring of at least 8 atoms.

5. What reactions are used to make alkenes? The ELIMINATION reaction we saw for the haloalkanes is one. There are two common types of elimination reactions. Elimination from haloalkanes is caused by treating the haloalkane starting material with a strong base, usually accompanied by heat. Elimination from alcohols is caused by treating the alcohol starting material with acid and heat. In many cases, these reactions generate a mixture of alkene products – there are several carbons with H's adjacent to the leaving group. The character of the mixture can be predicted by "Zaitsev's Rule". Again, the details of these reactions will be treated later. But it is good to know that these are common methods to prepare alkenes.

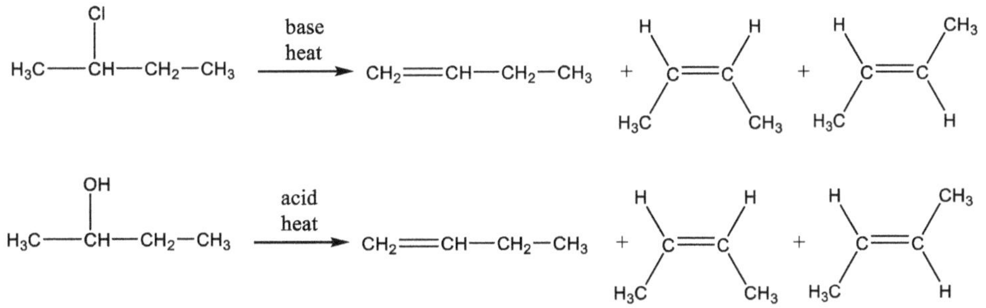

One additional preparation method is the Wittig (pronounced "Vi-tig") reaction. In this reaction, an aldehyde or ketone with a carbon-oxygen double bond is treated with a reagent known as a phosphonium ylid (a good word for Scrabble ™!) which effectively has a phosphorus-carbon double bond. The two reactants essentially exchange double bond partners, to produce a carbon-carbon double bond (the alkene) and a byproduct of triphenylphosphine oxide with a phosphorus-oxygen double bond. This reaction makes the alkene double bond in one quick step, and with less of an issue of multiple products, but the formation of the byproduct limits this reaction to a relatively small scale. You should know how to make alkenes.

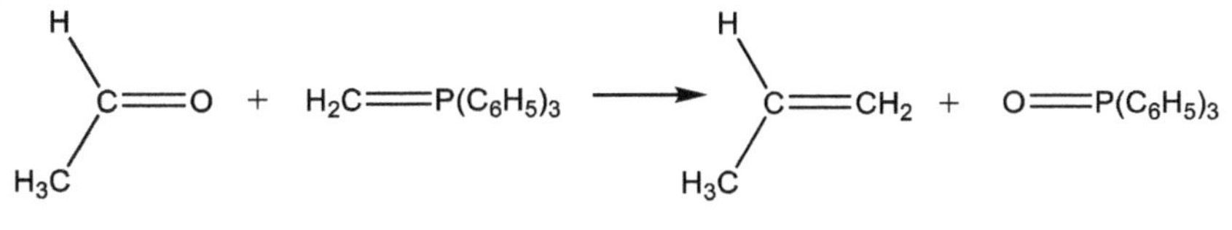

49

6. What reactions do alkenes do? The main type of reaction for alkenes is called ADDITION. In an addition reaction, a reagent with two components "X-Y" reacts with the alkene to connect one compenent to each carbon of the double bond by new single bonds – the double bond is lost, as is the bond connecting X and Y. There are a variety of reagents which can function as "X-Y" : H-H, H-Cl, H-Br, H-OH, Br-Br, Cl-Cl, and more.

$$H_2C{=}CH_2 \ + \ X{-}Y \ \longrightarrow \ \underset{CH_2-CH_2}{\overset{\overset{\displaystyle X \quad Y}{|\quad |}}{}}$$

In some reactions, the reagent that adds to the alkene is another alkene, then another. In this way, a long chain of carbons is produced. Such compounds are called POLYMERS (from the Greek words "poly" = many and "mer" = part or unit), and the reaction is called polymerization.

Some reagents that cause oxidation (Oxidants) will cause the double bond to break. These include ozone (O_3) and potassium permanganate ($KMnO_4$). Oxidation is written generically as [O].

$$\underset{R_1}{\overset{R_2}{C{=}C}} \ \xrightarrow{[O]} \ \underset{R_1}{C{=}O} \ + \ O{=}\underset{}{\overset{R_2}{C}}$$

Finally, there is a reaction where an alkene reacts with a compound with TWO double bonds (called a diene). The result is the formation of a six-membered ring with one double bond (a cyclohexene). This reaction is called the Diels-Alder reaction. This reaction is often treated briefly in sophomore organic chemistry, but treated much more fully in more advanced organic classes. We will do a lot more with addition later – but you will need to know these general types of reactions.

7. What are some uses and ideas related to the alkenes? The addition reaction makes alkenes great starting materials for making other compounds. The polymerization reaction is very important in industrial chemistry – this how polystyrene (=styrofoam for cups and coolers) and polyethylene (trash bags) and many, many others are made. There are a number of natural products that are alkenes – limonene, pinene, carotene, squalene. The cis/trans isomerism shows up in some of them: we already discussed how saturated fats are alkane type structures, while unsaturated fats are alkenes. All the naturally occurring unsaturated fats are *cis* alkenes. Unsaturated fats are usually liquid at room temperature and are then called OILS. If we do an addition reaction on unsaturated fats with H-H, we can convert the unsaturated to saturated: vegetable oil becomes solid vegetable shortening! If we do this reaction with a deficiency of H-H, we get "partially hydrogenated vegetable oil". Some double bonds remain, and in the process some of the *cis* double bonds become *trans* double bonds. These are then called "trans fats", unsaturated fats that behave like saturated, and considered to be unhealthy. You will often see food advertised as having "0 g trans fat" – but if the ingredients include "partially hydrogenated vegetable oil" there will be SOME trans fat – but the FDA allows them to round down to 0 if it is less than 0.5 g per serving!

Quick Review

1. ALKENES: Alkenes are compounds with a carbon-carbon double bond. They are unsaturated hydrocarbons, nonpolar, with low MP and BP, insoluble in water.

2. NOMENCLATURE: The alkenes are named with the –ene suffix. The chain is numbered to give the double bond the lower numbers, and the position indicated by the lower of the two double bond numbers: 2-butene, for example. Substituents are indicated by name and number: 2-methyl-2-butene.

3. STEREOCHEMISTRY (CIS/TRANS): Alkenes do not rotate around the double bond. If both double bond carbons have 2 different groups attached, then there can be cis-trans isomers. The cis-trans nomenclature can only be used if there are identical groups on the 2 double bond carbons. The IUPAC has devised a nomenclature system (E and Z) that can be used in all possible cases.

4. CYCLOALKENES: Rings containing the double bond are cycloalkenes. The ring is always numbered so the double bond carbons are 1 and 2. The cycloalkenes can be more strained than the cycloalkanes.

5. PREPARATION: Alkenes form by elimination from haloalkanes or alcohols, or by the Wittig reaction.

6. REACTIONS: The most common reaction of alkenes is addition. Many reagents add to the alkene double bond. Alkenes also react by polymerization, oxidative cleavage, and the Diels-Alder reaction.

7. USES: Alkenes are used as starting materials for many reactions, although polymerization is most important in industry. Alkanes and alkenes are found in fats and oils, and we recognize that these can be saturated or unsaturated, and the unsaturated are naturally *cis* but can be made *trans*.

10.   Alkynes

1. Alkynes are the hydrocarbons that have a triple bond. The carbon-carbon triple bond is the functional group of the alkynes. The triple bond carbons are sp hybridized – they form a carbon-carbon sigma bond with the sp hybrid orbitals, then two pi bonds with their remaining p orbitals. The two pi bonds are at $90°$ to each other. The geometry around sp hybridized carbons is linear – $180°$ – and with only one group attached to each triple bonded carbon, there are no cis-trans stereoisomers. As hydrocarbons, the alkynes are nonpolar. So they have low MP, low BP, are insoluble in water, low densities and all the other properties similar to the alkanes and alkenes.

ALKYNE PHYSICAL PROPERTIES

Nonpolar
Low MP/BP
Insoluble in water
Low densities

2. The IUPAC nomenclature of the alkynes models that of the alkenes. The suffix that indicates the triple bond is –yne. These are the rules for naming alkynes: RULE 1. Locate the longest chain that includes both carbons of the triple bond. This is the parent chain. The name of the alkyne will be based on the alkane of this length with the suffix changed from –ane to –yne. So a two carbon chain with a double bond is ethyne. RULE 2. Number the chain so the triple bond has the lower set of numbers. There are 2 carbons for the triple bond, so there are 2 numbers associated with the triple bond. In some cases, the triple bond will be in the center of the chain, so it will not make any difference which way we number. RULE 3. Use the lower of the two numbers to indicate the location of the triple bond. So if there is a 4 carbon chain with the triple bond at one end, the triple bond is between carbons 1 and 2. We use the lower number – 1 – to indicate the location: this is 1-butyne. If there is a 4 carbon chain with the triple bond in the middle, it is between carbons 2 and 3, and we use "2": this is 2-butyne. RULE 4. If there are substituents, list them by name and number before the rootname, the same as the alkanes. For those alkynes where the double bond is in the center, if there are substituents, number to give the lower number at the first point of difference (just like the alkanes). Use this numbering for RULE 3. These rules allow us to name the alkynes. There is one alkyne with an old, trivial name: the two carbon alkyne, ethyne, is usually called by its old name of acetylene. Always good to know nomenclature!

$$HC\equiv CH-CH_2-CH_3 \qquad\qquad H_3C-CH_2-C\equiv C-\underset{\underset{CH_3}{|}}{CH}-CH_3$$

1-butyne   2-methyl-3-hexyne

3. Alkynes can be incorporated into rings, but cycloalkynes are rather uncommon. Since the sp carbons of the alkyne are linear, there are four carbons that make a straight line. It takes at least 4 other carbons to connect the ring, and 6 with reduced strain, so cyloalkynes smaller than cyclooctyne do not exist. You are unlikely to ever see a cycloalkyne.

4. How do we make the alkynes? It is possible to do ELIMINATION reactions to make alkynes. We start with an alkene that has a hydrogen on one double bond carbon and a bromine on the other. Treatment with a strong base (usually an amide base like sodium amide ($NaNH_2$) or lithium di-isopropylamide "LDA") causes loss of H and

Br to make the triple bond. This also works on alkanes with two bromines on either the same carbon or adjacent carbons.

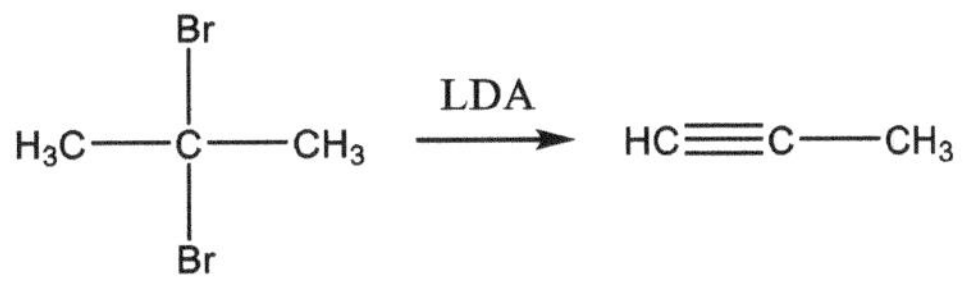

Another way to make alkynes is to start with an existing "terminal" alkyne – that is, where the alkyne is at the end of the chain (ethyne, propyne, 1-butyne, 1-pentyne, etc.) It is found that the hydrogen attached to the triple bond is relatively acidic: treatment with a strong base can remove this hydrogen and make an alkyne anion (an "acetylide" anion). These can be used as nucleophiles in NUCLEOPHILIC SUBSTITUTION – the alkyne anion displaces the halogen of a haloalkane, and the two carbon pieces (alkyne and haloalkane) are coupled together to make the new larger alkyne. The reason for this acidity is the sp hybridization: sp^3 and sp^2 hybrids have more high energy p character and have pK_a values over 40, but sp hybrids with their greater low energy s character have pK_a values around 25. Alkyne acidity and using the anions as nucleophiles is a common question.

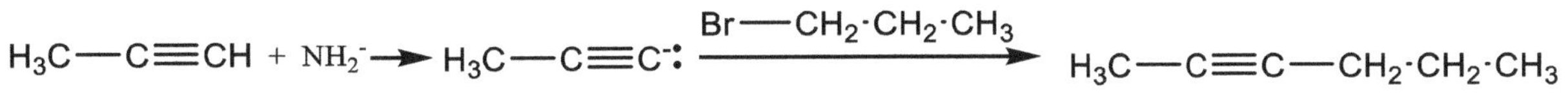

5. What are the reactions of the alkynes? We just did a few of them! Alkynes undergo acid-base reactions if the alkyne is at the end of the chain. The resulting acetylide anions can act as nucleophiles in nucleophilic substitution.

Alkynes also can react by ADDITION like the alkenes. Since there are two pi bonds, there can be two rounds of addition! After one round of addition, the alkyne becomes an alkene. With two rounds of addition, the alkyne becomes an alkane. In some single additions, different reagents produce a cis or a trans product. Addition of water leads to alcohols on the double bond – known as enols – which are unstable and rearrange to the carbonyl compound (aldehyde or ketone). The generic example below doesn't show all of the detail, but is useful for remembering the general reaction.

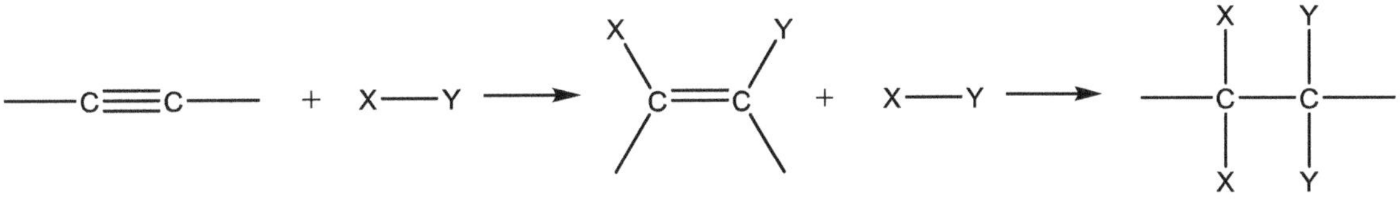

6. What are some uses of alkynes? They are useful as starting materials. Addition of H_2 can make cis and trans alkenes; the coupling reaction with acetylene anions is a useful way to put together new, larger hydrocarbon frameworks from smaller pieces. Finally, ethyne or acetylene burns with an extraordinarily hot flame. So an acetylene torch is used to cut steel.

Only 6 points!

Quick Review.

1. ALKYNES: Alkynes are compounds with carbon-carbon triple bonds.
2. NOMENCLATURE: Alkyne nomenclature is like that of alkenes: the suffix –yne indicates the triple bond, a number indicates the position in the chain.
3. CYCLOALKYNES: Cyloalkynes smaller than 8 carbons do not exist; in sophomore organic chemistry cycloalkynes are rare.
4. PREPARATION: Alkynes can be made by elimination or by substitution starting with a terminal alkyne.
5. REACTIONS: Alkyne reactions: The hydrogen at the end of a terminal alkyne can be removed by strong bases. The alkyne (acetylide) anions act as nucleophiles in nucleophilic substitution. Alkynes undergo addition with the same reagents as alkenes, but can usually add two equivalents of the reagent. Addition of water leads to enols that rearrange to carbonyl compounds.
6. USES: Alkynes are useful in the preparation of other compounds. Ethyne or acetylene is used as a fuel.

11.   Alcohols

1. Alcohols are those compounds that have a hydroxyl group (-OH: oxygen with a connected hydrogen) as the functional group. The hydroxyl group can be seen as a piece of a water molecule – with an alkyl substituent replacing one of water's hydrogens. The hydroxyl group has properties related to those of water – it is polar, including hydrogen bonding between the hydrogen of one alcohol and the oxygen of another. The polarity gives alcohols high MP and BP. It also makes the smaller alcohols soluble in water – alcohols are considered to be soluble up to 4 carbons. (This means that 4 carbon alcohols – the butanols – dissolve at more than 10 g/100 mL of water, it is right at the threshold for the lowest. The next largest alcohols – the pentanols – mostly dissolve less than 10 g/100 mL, ~2 for most isomers – but they are considered insoluble!) Alcohols of 1, 2, and 3 carbon chains are miscible in water – that means they mix with water in any proportions. The density is higher than the alkanes, but lower than water.

Alcohol Properties

Polar
High BP and MP
Water soluble to C4
Moderate density

2. In the IUPAC nomenclature, the presence of the hydroxyl functional group is indicated by the –ol suffix. So methane becomes methanol, ethane becomes ethanol. The rules for naming alcohols are as follows: RULE 1. Identify the longest chain with the hydroxyl attached. The corresponding alkane is the parent structure, and the alcohol is named by adding the suffix –ol in place of the –e of the alkane name. (propanol, butanol, etc.) RULE 2. Number the chain to give the hydroxyl the lower number. If the hydroxyl is in the middle of the chain, number the chain to give any other substituents the lower number at the first point of difference. RULE 3. Indicate the position of the hydroxyl by number: 1-propanol has the hydroxyl on an end carbon, 2-propanol has the hydroxyl in the middle, on the 2nd carbon. RULE 4. Indicate any other substituents by name and number: 2-methyl-3-pentanol, etc. With these rules, you can name all linear alcohols. To add cyclic alcohols to the list, we do the following: Name the ring as the cycloalkane. Convert to the cycloalkanol with the –ol suffix. Cyclic alcohols are always numbered to give the alcohol carbon # 1. So no number is necessary: cyclopentanol rather than 1-cyclopentanol. Numbers go around the ring in whatever direction gives the lower number at the first point of difference: 3-methylcyclopentanol, etc. There can be *cis* and *trans* isomers, of course!

2-butanol   2,5-dimethyl-4-heptanol   2,2-dimethylcyclobutanol

3. How do we make alcohols? There are 3 common ways: ADDITION to alkenes, REDUCTION of carbonyls, and nucleophilic substitution. Addition to alkenes can be accomplished in several ways. Acid catalyzed hydration, where the elements of water H and OH add to the alkene double bond. As with addition of HX to the alkenes (see above), the product of this can be predicted by Markovnikov's Rule. This is a simple reaction, but there are certain

problems, so alternative methods have been developed. Sometimes the carbon skeleton can change during the reaction ("rearrangement"), and sometimes the desired product is not the Markovnikov product. The solution to rearangement is called Oxymercuration-Reduction. The alkene is treated with mercury (II) acetate ($Hg(OAc)_2$) in a mixture of water and the solvent tetrahydrofuran (THF), and the alkene adds OH from water and Hg from the salt. The Hg is removed by treatment with sodium borohydride ($NaBH_4$), a reduction, which replaces it with H. The net result is addition of H and OH – just like addition of H_2O. This gives the Markovnikov product, but does not rearrange. To make the other product (Anti-Markovnikov, it is called), the method is called Hydroboration-Oxidation. The alkene is treated with borane, (BH_3) which adds B and H across the double bond. The B is converted to an OH with hydrogen peroxide (H_2O_2), an oxidation. This adds H and OH, with no rearrangement, and gives the opposite product to the Markovnikov reactions. The control we get from these three variations makes addition to alkenes the most popular way to make alcohols.

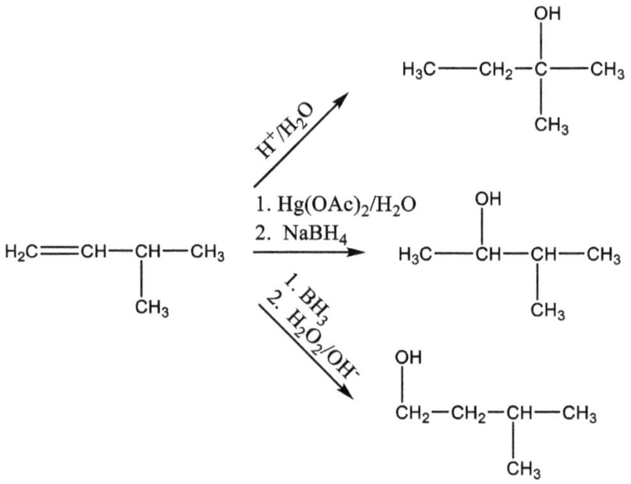

REDUCTION of carbonyls is also convenient. Aldehydes and ketones can be reduced with sodium borohydride ($NaBH_4$) to make the alcohols – we essentially add H to both the C and O of the carbon-oxygen double bond, so the carbonyl O becomes an OH. We can also reduce carboxylic acids, but it requires a stronger reducing agent: lithium aluminum hydride ($LiAlH_4$).

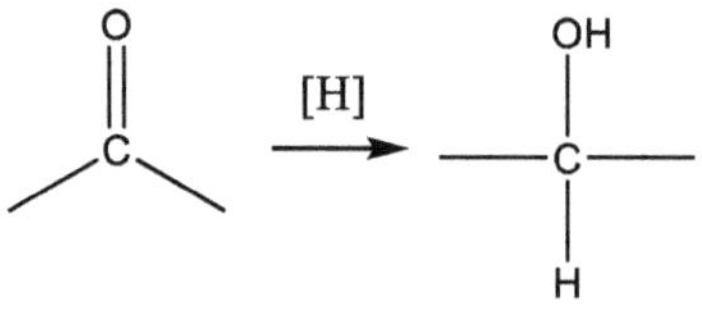

Finally, we can use NUCLEOPHILIC SUBSTITUTION: treatment of haloalkanes with water or hydroxide replaces the halogen with a hydroxyl. Because the haloalkanes are most commonly made from the alkenes, there is little advantage to this chemistry.

4. What reactions do alcohols do? Elimination to make alkenes, dimerization to make ethers, substitution to make haloalkanes, nucleophilic reactions to make ethers and ester, oxidation to carbonyl compounds. Alcohols are also very weak acids – too weak to act as acids in water, but they can donate a proton. Alcohols can be DEPROTONATED as acids by reaction with strong bases or (more commonly) with alkali metals (Na, K). The resulting alkoxide anions are strong bases and nucleophiles.

$$R-OH + Na \longrightarrow R-O^-Na^+ + H_2$$

ELIMINATION occurs when we treat an alcohol with acid and high temperatures. The exact strength of acid and temperature depends on the alcohol – primary alcohols need stronger conditions, tertiary alcohols need milder conditions.

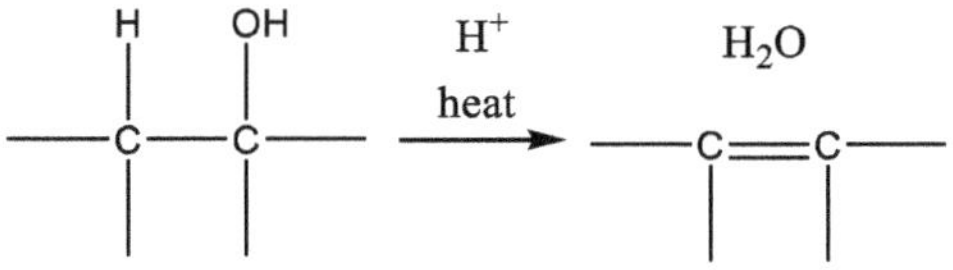

$$\overset{H}{\underset{\vert}{\overset{\vert}{C}}}-\overset{OH}{\underset{\vert}{\overset{\vert}{C}}} \quad \xrightarrow[\text{heat}]{H^+} \quad \overset{H_2O}{} \quad C=C$$

DIMERIZATION takes two alcohols and combines them with loss of water: so two ethanols make an ether with two ethyl groups, two propanols make and ether with two propyl groups, and so on. The tricky thing is that dimerization requires conditions only a little different than elimination: acid and moderately high temperatures.

$$R-OH + HO-R \quad \xrightarrow[\text{heat}]{H^+} \quad \overset{H_2O}{} \quad R-O-R$$

SUBSTITUTION requires special reagents – we mentioned these up in the haloalkane section. Thionyl chloride converts alcohols to chloroalkane, and phosphorus tribromide converts alcohols to bromoalkane.

$$R-OH \xrightarrow{SOCl_2} R-Cl \qquad R-OH \xrightarrow{PBr_3} R-Br$$

More commonly the hydroxyl acts as a NUCLEOPHILE. In these reactions, the alcohol hydroxyl displaces a halogen. If the halogen is on a haloalkane, then two alkyl groups are now connected by an oxygen – an ether. One alkyl group comes from the alcohol, and one from the haloalkane. If the halogen is connected to a carbonyl (a carbonyl halide), then the product is an ester.

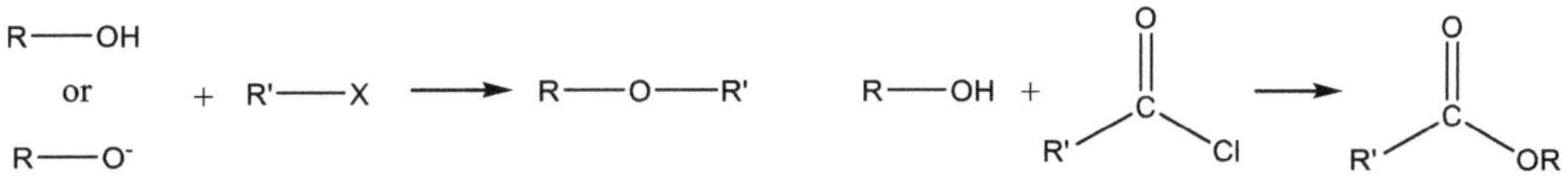

$$\begin{array}{c} R-OH \\ \text{or} \\ R-O^- \end{array} + R'-X \longrightarrow R-O-R' \qquad R-OH + \underset{R'}{\overset{O}{\overset{\|}{C}}}-Cl \longrightarrow \underset{R'}{\overset{O}{\overset{\|}{C}}}-OR$$

OXIDATION is the reverse of the reductions we saw above. It is a little more complex – there are a wider variety of oxidizing agents: Sodium Dichromate ($Na_2Cr_2O_7$) was once the oxidant of choice, but Chromium is toxic. Potassium permanganate ($KMnO_4$) is still widely used, but is stronger than needed. Sodium hypochlorite (NaOCl) or bleach is inexpensive, but not always sufficiently strong. Treatment of secondary alcohols with these reagents gives ketones as a product. Primary alcohols are more difficult. Treatment of primary alcohols with these reagents oxidizes to the carboxylic acids. Because these reagents are aqueous, it is impossible to stop at the intermediate aldehyde stage. To make the aldehyde, we must oxidize in an anhydrous organic solvent: a common way is to use pyridinium chlorochromate (PCC) in dichloromethane. Where appropriate, details of these reactions will be covered later. For this section what we want to learn are the multiple transformations where alcohols are the starting point.

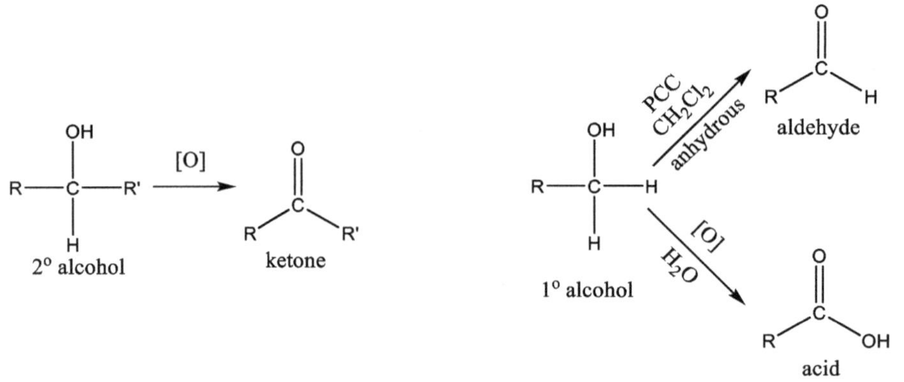

OH
[O]
R—C—R' ———→ 2° alcohol → ketone

OH
R—C—H
H
1° alcohol

PCC / CH$_2$Cl$_2$ anhydrous → aldehyde

[O] / H$_2$O → acid

5. One special preparation from alcohols that is often described is the Williamson Ether Synthesis. This set of reaction requires knowledge of several of the reaction shown above, so your professor is likely to use the Williamson Ether Synthesis on a test. This may take the form of asking you to propose how to make an ether using the Synthesis, explaining the reactions, or some other combination.

In the Williamson Ether Synthesis, one alcohol is the nucleophile and a second is the substrate for nucleophilic substitution. When they combine, an ether is formed. If the alcohols are different (they usually are), it is an unsymmetrical ether. (Remember from point 4, symmetrical ethers can be made by acid catalyzed dehydration at moderate temperature.) To make the one alcohol into a strong nucleophile, it is deprotonated to the alkoxide with sodium metal (Na) – the first reaction in point 4 above. The corresponding anions are named from the alcohol: methoxide, ethoxide, 2-methyl-3-pentoxide. The alkoxide bases are stronger than hydroxide (we see this in the pKa values: water is 15.6, while the alcohols range from 16 to 18). It is possible to use a stronger base, such as sodium amide (NaNH$_2$), but the common way is to treat the alcohol with sodium metal. This makes one alcohol into a strong nucleophile. The other alcohol is converted into a good substrate for nucleophilic substitution. This can be done by making the haloalkane, using thionyl chloride or phosphorus tribromide as described in point 4 above. A sometimes better way is to make the alkyl sulfonate. Here the alcohol reacts with a sulfonyl chloride (most commonly methanesulfonyl chloride or p-toluenesulfonyl ("tosyl") chloride as shown), displacing the chloride with the alcohol oxygen. The product is an alkyl sulfonate ester – similar to the ester reaction shown in point 4 above, but we have replaced the carbonyl chloride / carboxylate ester with the sulfonyl chloride / sulfonate ester. Because the sulfonate anion is a very weak base (the sulfonic acids are almost as strong as sulfuric acid), it is a good leaving group that leaves as easily as chloride or bromide. The advantage is that this reaction does not break any bonds to carbon, so any stereochemistry is always retained! (Remember – that means it does not change.)

So now we have taken two alcohols and one has become the nucleophile (alkoxide) and one the substrate (bromide, chloride, or sulfonate). If we combine them in a good solvent for the nucleophilic substitution reaction (N,N-dimethylformamide or DMF is a common choice), then we couple the two pieces together to make an ether. We have to choose which alcohol is which: which to make into the alkoxide and which to make into the substrate. This is actually easy to do: the alcohol with fewer alkyl substituents is the better substrate, the alcohol with more alkyl substituents is the better nucleophile (alkoxide). So for the alkoxide, 3° over 2° over 1°; while for the substrate 1° over 2° over 3°. The stereochemistry of the reactions can all be controlled: alkoxide formation and sulfonate ester formation both occur with RETENTION of stereochemistry, while the S$_N$2 reaction occurs with INVERSION of stereochemistry. So if we know what structure or isomer we want at the end, we can choose which alcohols (and which isomers) to start with. In the following example, we choose the 1° alcohol to be substrate, and the 2° alcohol to be the alkoxide. Since the alkoxide is formed and reacts with RETENTION, we use the (S)-alcohol to make the (S) product.

$$H_3C\text{——}CH_2\text{—}O\text{——}\overset{\displaystyle CH_3}{\underset{\displaystyle CH_2CH_3}{C}}\text{''''''}H$$

Target Molecule:
(S)-2-ethoxybutane

1° alcohol as substrate: methanesulfonate leaving group

$$H_3C\text{——}CH_2\text{—}OH + H_3C\text{——}\overset{O}{\underset{O}{\overset{\|}{\underset{\|}{S}}}}\text{—}Cl \xrightarrow{\text{pyridine}} H_3C\text{——}CH_2\text{—}O\text{——}\overset{O}{\underset{O}{\overset{\|}{\underset{\|}{S}}}}\text{—}CH_3$$

2° alcohol as alkoxide nucleophile

$$HO\text{——}\overset{\displaystyle CH_3}{\underset{\displaystyle CH_2CH_3}{C}}\text{''''''}H + Na \longrightarrow Na^+ \; {}^-O\text{——}\overset{\displaystyle CH_3}{\underset{\displaystyle CH_2CH_3}{C}}\text{''''''}H + H_2$$

The (S) alcohol gives the (S) alkoxide and the (S) ether product" RETENTION

Substrate and nucleophile give product: target molecule

$$H_3C\text{——}CH_2\text{—}O\text{——}\overset{O}{\underset{O}{\overset{\|}{\underset{\|}{S}}}}\text{—}CH_3 + Na^+ \; {}^-O\text{——}\overset{\displaystyle CH_3}{\underset{\displaystyle CH_2CH_3}{C}}\text{''''''}H \longrightarrow H_3C\text{——}CH_2\text{—}O\text{——}\overset{\displaystyle CH_3}{\underset{\displaystyle CH_2CH_3}{C}}\text{''''''}H + CH_3SO_3^- \; Na^+$$

If a chiral 2° alcohol is the substrate, the reaction proceeds with INVERSION and we need to start with the (R) alcohol to make the (S) product. In the next example, the 3° alcohol is the alkoxide and the 2° is the substrate, so we use the (R) substrate to wind up with the (S) product.

Williamson ether syntheses are favorites of professors – when you are studying this topic, be sure to understand it!

$$H_3C\text{——}\overset{\displaystyle CH_3}{\underset{\displaystyle CH_3}{C}}\text{—}O^-Na^+ + \overset{\displaystyle H_3C}{\underset{\displaystyle H_3CH_2C}{C}}\text{''''''}{}_H\text{—}OSO_2CH_3 \longrightarrow H_3C\text{——}\overset{\displaystyle CH_3}{\underset{\displaystyle CH_3}{C}}\text{—}O\text{——}\overset{\displaystyle CH_3}{\underset{\displaystyle CH_2CH_3}{C}}\text{''''''}H + CH_3SO_3^- \; Na^+$$

6. What are the uses of alcohols? As is obvious from the last section, alcohols are the beginning point for many other preparations: alkenes, ethers, esters, aldehydes, ketones, carboxylic acids… Alcohols are also used as solvents for many organic substances, whether for reactions or paints and coatings. Alcohols occur in a variety of naturally occurring substances, including as products of fermentation: ethanol is produced by yeasts and is then found in the alcoholic beverages such as wine and beer. Alcohols are generally toxic however, and at some point the yeast are killed by their own product. This toxicity is also useful – isopropyl alcohol or 2-propanol (rubbing alcohol) is used to clean wounds and skin, ethanol with a gelling agent is used as hand sanitizer. Higher concentrations of alcohol in beverages must be produced by distillation. Alcohol is also toxic to humans. Certain enzymes in the liver can detoxify ethanol (but not odd numbered alcohols like methanol or propanol!), but if more alcohol is consumed than the liver can process, death can result – ethanol poisoning occurs frequently in college aged *homo sapiens* – don't let it be you!

Quick Review

1. ALCOHOLS: Alcohols have a hydroxyl group. They are polar and have higher boiling/melting points and greater water solubility than the hydrocarbons.

2. NOMENCLATURE: The suffix that designates an alcohol is –ol. The hydroxyl group is always given the lower number, like 2-methyl-3-pentanol.

3. PREPARATION: Preparation of alcohols is accomplished primarily by addition to alkenes, reduction of carbonyls, or nucleophilic substitution.

4. REACTIONS: Alcohols react in a number of ways. Elimination to alkenes, oxidation to make aldehydes, ketones, and carboxylic acids, as nucleophiles to make ethers and esters. Alcohols can function as acids or bases. Deprotonation makes the alkoxide (which is a stronger nucleophile than the alcohol), protonation of the hydroxyl makes a better leaving group (H_2O) for elimination and nucleophilic substitution.

5. WILLIAMSON ETHER SYNTHESIS: The Williamson Ether synthesis is a method for making unsymmetrical ethers from two alcohols.

6. USES: Alcohols have many uses: Fuels, disinfectants, solvents, reagents and starting materials for other reactions.

12.   Ethers

1. Ethers are compounds where two organic groups are linked by an oxygen. We can think of a progression from water (H-O-H) to alcohols (R-O-H) to ethers (R-O-R, although the two R groups can be different, so we often write R-O-R'). The most common ether has two ethyl groups: "diethyl ether" –this is sometimes just called "ether". Although the C-O bonds are polar, and the oxygen has lone pairs that can participate in hydrogen bonding, ethers are mostly non-polar. They have low boiling and melting points: usually only a few degrees different from the alkane of similar structure and MW. (So pentane CH_3-CH_2-CH_2-CH_2-CH_3 and diethyl ether CH_3-CH_2-O-CH_2-CH_3 differ in boiling point by only a few degrees.) Ethers of up to three carbons are soluble in water, since the oxygen can accept hydrogen bonds from water. The solubility of ethers tends to be higher in acidic solutions of water – so we usually try not to use diethyl ether for extractions of acidic aqueous solutions.

Ether Properties

Nonpolar
Low BP and MP
Water soluble to C3
Low density

2. The IUPAC names ethers as substituted alkanes. The longer of the two alkyl groups becomes the parent alkane chain, the other with the oxygen is treated as a substituent, an "alkoxy" group. So $-O-CH_3$ is methoxy, $-O-CH_2-CH_3$ is ethoxy, $-O-CH_2-CH_2-CH_3$ is propoxy and so on for butoxy, pentoxy, etc. The name of the alkyl group is changed to alkoxy to indicate it is connected through an oxygen. Many simpler ethers have "trivial" names from an older nomenclature system, where the two groups are listed followed by the word "ether". So CH_3-O-CH_3 is methoxymethane in the IUPAC system (since the two groups are the same length, either can be parent and either substituent), dimethyl ether in the old system; $CH_3-CH_2-O-CH_3$ is 1-methoxyethane in IUPAC, methyl ethyl ether in the old system; and finally $CH_3-CH_2-O-CH_2-CH_3$ is 1-ethoxyethane in IUPAC, diethyl ether in the old system. A lot of chemists still use the old system for simple ethers (diethyl ether is almost never referred to by its IUPAC name), either out of habit or because the names are more descriptive or easier to say (they aren't really shorter). But when the alkyl groups become complex, the old system becomes very tedious to use and everyone uses the IUPAC system. So the structure below is easily named by the IUPAC system as 3-ethyl-1-methoxy-2,4-dimethylpentane. The old nomenclature has to generate a name for the branched group, which is fairly cumbersome.

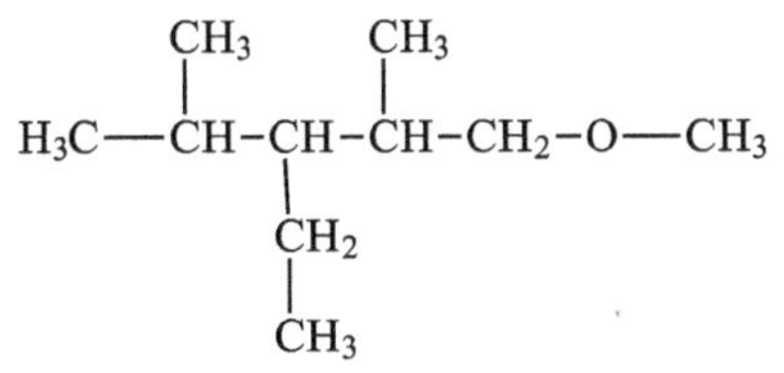

$$H_3C-\underset{\underset{\underset{CH_3}{|}}{\underset{CH_2}{|}}}{CH}-\underset{\underset{CH_3}{|}}{CH}-\underset{\underset{CH_3}{|}}{CH}-CH_2-O-CH_3$$

3-ethyl-1-methoxy-2,4-dimethylpentane

Ethers can also be cyclic – one alkyl chain connects at each end to the oxygen. These are the first examples we've seen of "heterocycles" – ring structures incorporating a non-carbon atom ("heteroatom"). We will see that trivial names tend to be very common with heterocycles, and this is true of the cyclic ethers. The IUPAC and commonly used trivial names for the most common cyclic ethers are given below (there is no common name for the four membered ring ether). In theory, the IUPAC names are "always correct", but the five and six membered cyclic ethers are very rarely referred to by the IUPAC names, and most commonly by their abbreviated common names "THF" and "THP". THF and dioxane are very common solvents in organic chemistry, and ethylene oxide is a common industrial chemistry (but rarely used in research laboratories due to its hazardous nature). The four membered ring ether is not common, so there is no common name. In a general sense, the prefix "oxa" can be used to designate replacement of a carbon with an oxygen, so "oxacyclodecane" would be an acceptable name for a ten membered ring cyclic ether. The 1,4-dioxane name for the 6-membered, two oxygen ring is a shortening of the more formal "1,4-dioxacyclohexane". Always learn the nomenclature!

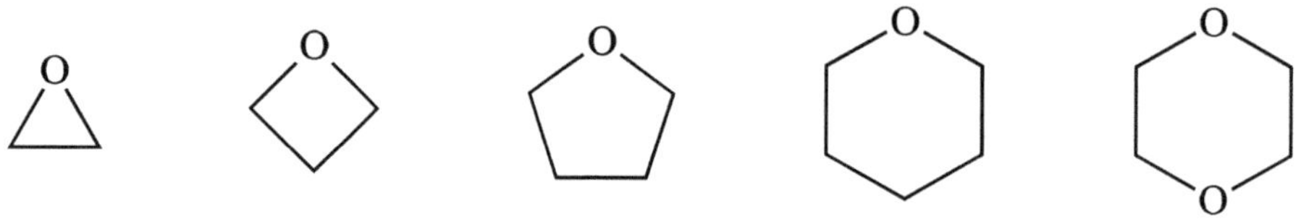

IUPAC	oxirane	oxetane	oxolane	oxane	1,4-dioxane
common	ethylene oxide	None	tetrahydrofuran "THF"	tetrahydropyran "THP"	dioxane

3. Ethers are most often prepared from alcohols. Symmetric ethers can be prepared by heating an alcohol with acid to a temperature slightly lower than is required to do elimination. So ethanol, treated with concentrated sulfuric acid and heated to ~140 °C will give mainly diethyl ether. Heating to a higher temperature (~180 °C) will give mainly the alkene - ethene. The exact temperatures required depend on the alcohol structure – 1° alcohols require higher temperatures, 2° lower temperatures, and 3° the lowest. The mechanism of this acid catalyzed dimerization of alcohols is shown below – but we will get into more mechanism details in the next part. Unsymmetric ethers are often prepared by the Williamson Ether Synthesis described above in the alcohol section.

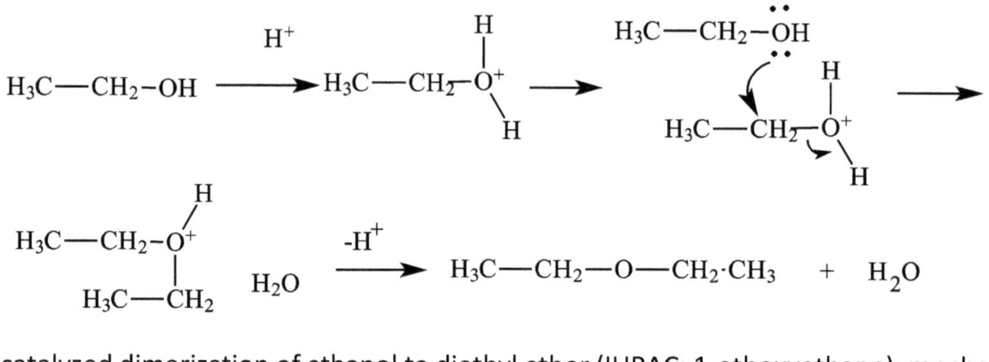

Acid catalyzed dimerization of ethanol to diethyl ether (IUPAC: 1-ethoxyethane): mechanism

4. Ethers are relatively unreactive. As we will see in the next section, that is one reason why they tend to make good solvents. The most common reaction of ethers is acid catalyzed cleavage – almost the reverse of the dimerization shown above. "Almost" because the reagent that is used is HBr instead of H^+ and H_2O: Br^- is a better nucleophile than H_2O, so it proceeds more rapidly.

R-O-R + HBr → R-OH + R-Br

The ether is protonated on the oxygen (that is the reverse of the final step of the dimerization), then Br⁻ acts as a nucleophile to cleave the ether – Br⁻ being a better nucleophile than H_2O. The Br- always attacks the more susceptible group: methyl > 1° > 2°> 3° or aryl. So in cleavage of methyl t-butyl ether (IUPAC: 2-methoxy-2-methylpropane), the products are bromomethane and 2-methyl-2-butanol: the bromide anion reacts at the less hindered methyl not the more hindered tertiary center of the t-butyl group.

Cleavage of methyl t-butyl ether (IUPAC: 2-methoxy-2-methylpropane) by HBr

Preparations and reactions of ethers may show up as questions – as noted above, the Williamson Ether Synthesis allows professors to see if you understand the options and the mechanisms for a variety of reactions. Other ether preparations and reactions might be chosen from the menu of reactions available for questions.

5. As noted above, one of the main uses of ethers is as solvents. Because the ethers have some polar character – with the polar C-O bonds and the Oxygen lone pairs – they can dissolve more polar reagents; but the nonpolar part of the ether is good at dissolving nonpolar organic substances. Since the ethers themselves are relatively unreactive, this allows us to dissolve many different things in ether solvents to perform many reactions. Some of the common ether solvents are diethyl ether; the 5 membered ring THF; and the 6 membered ring with two oxygens, dioxane; but there are many more (dimethoxyethane or "glyme", bis(2-methoxyethyl) ether or "diglyme", and a whole series of ethers of ethylene glycol (2,3-ethanediol) and propylene glycol (1,3-propanediol) called the "glycol ethers").

Diethyl ether was once used in medicine as an anesthetic. "To be put under the ether" is sometimes still used to mean use of a general anesthetic. But this use was given up many years ago because of the volatility and flammability of diethyl ether, which made operating rooms very susceptible to fires. In modern medicine, it is now rare to use an inhaled general anesthetic (administration through an IV is more common since it gives better dose control and more rapid action), but when an inhaled anesthetic is used, it is usually a non-flammable haloalkane.

Certain ethers are particularly good at dissolving particular cations. These are the "crown ethers", which have repeating segments of –CH_2CH_2O- around a ring. So 5 segments has 15 atoms and 5 oxygens and is called "15-crown-5": the inside of the ring is the right size for a sodium cation Na^+ to be supported by lone pairs of electrons from all 5 oxygens. The next size up has 6 segments (18-crown-6) and has a ring that is the right size for potassium ions K^+. Adding a small amount of 15-crown-5 will enhance the solubility of sodium salts in organic solvents, while 18-crown-6 enhances the solubility of potassium salts.

Quick Review

1. ETHERS: Ethers have two alkyl groups attached to an oxygen. They are nonpolar.

2. NOMENCLATURE: IUPAC names ethers as alkoxyalkanes. The cyclic ethers have special names.

3. PREPARATION: Ethers are prepared from alcohols – symmetric by dimerization, unsymmetric by Williamson Ether Synthesis.

4. REACTIONS: Ethers are fairly unreactive – but can be cleaved by treatment with HBr

5. USES: Ethers are used as solvents. Diethyl ether was used as an anesthetic. Crown ethers form complexes with cations.

IV. Reaction Mechanisms

13.   Nucleophilic Substitution

1. We have already seen many nucleophilic substitutions – the Williamson Ether synthesis and cleavage of ethers by HBr. In these reactions, a nucleophile (Nu) replaces a leaving group (L) on the carbon compound (R-L) or substrate, this becomes R-Nu, the product. Now we will examine this reaction in depth.

$$Nu^- \ + \ R\!-\!L \ \longrightarrow \ R\!-\!Nu \ + \ L^-$$

Nucleophilic substitution: Nucleophile Nu replaces leaving group L.

The NUCLEOPHILE is a reagent which has a lone pair of electrons, and wants to donate these electrons to a positive charge. The lone pair loves ("*philos*") nuclei (where positive charge is), therefore "nucleophile". Nucleophiles are Lewis Bases, but their strength is measured differently. Acids and Bases (including Lewis Bases) are evaluated based on the position of an *equilibrium*. Nucleophiles are evaluated on the *rates* of reactions – how fast they react. Faster reaction = stronger nucleophile. The leaving group is a group that can take a pair of electrons and leave. We can predict the strength of nucleophiles and leaving groups. Leaving groups are easy: we look at the conjugate acid of the leaving group. The stronger the conjugate acid, the better the leaving group – if a group will give up a bond to H^+ it will also give up a bond to carbon (R^+). Nucleophiles are a little harder to grasp – there are more nuances. Bigger atoms tend to be more nucleophilic – they have larger, longer electron pairs and it is easier for them to donate those pairs. So $I^- > Br^- > Cl^- > F^-$. Less electronegative atoms don't want the lone pair of electrons – so they are more nucleophilic: $C:^- > N:^- > O:^- > F:^-$. For the same nucleophilic atom, the stronger base is the better nucleophile. So anions are usually better than neutrals: OH^- better than H_2O, but also RO^- better than OH^- (since the pK_a of H_2O is 15.6 but alcohols are 16-18... remember stronger acids (lower pK_a) have weaker conjugate bases – RO^- are stronger bases and stronger nucleophiles than OH^-).

2. Nucleophilic substitution is where the concept of REACTION MECHANISM is usually introduced in organic chemistry. (I teased you with a couple of them in the ether section.) This is an important concept. Your instructor will expect you to understand (or at least regurgitate) mechanisms. The mechanism is the sequence of simple steps by which a reaction occurs. A reaction mechanism may have one step or a dozen. Since we can't actually observe the tiny atoms as they go through the steps, we can't prove that a given mechanism occurs. We can disprove mechanisms – if a certain mechanism predicts a certain outcome, and that outcome is not observed, then the mechanism CAN NOT be correct- it is disproven. So we think up all the possible mechanisms we can, then set about trying to disprove them. The one that can't be disproven is likely to be correct – unless there is a mechanism we didn't consider! So for nucleophilic substitution, we can consider 3 simple mechanisms: 1) break the bond to the leaving group in a first step to form a carbocation and the leaving group anion, then make the bond to the nucleophile in a second step; 2) break the bond to the leaving group and make the bond to the nucleophile simultaneously (all in one step); 3) make the bond to the nucleophile in one step, then break the bond to the leaving group in a second step. (We will only go over this procedure for this first mechanism – usually we will just provide the accepted mechanism – that is all you need to know!)

In mechanism (1), the slow step is breaking the bond. So the RATE of the reaction will only depend on the concentration of the substrate R-L, not on the concentration of the nucleophile: Rate = k[R-L]. This is called a UNIMOLECULAR or FIRST ORDER rate – it depends on only one concentration. The reaction proceeds through a

REACTIVE INTERMEDIATE – a CARBOCATION. A reactive intermediate is a high energy structure that occurs as part of the reaction. Carbocations are positively charged intermediates, where one carbon only has three bonds – where the fourth bond used to be there is an empty p orbital. Carbocations are planar and therefore achiral. Any products that come from a carbocation must be racemic. Carbocations can REARRANGE: they can move groups around to make a more stable cation. Groups on carbons next to the empty p orbital can shift over to make a bond to that carbon – this can be an H or an alkyl group. The shifting group brings 2 electrons with it so the empty orbital is now on the neighboring carbon. If this converts a 2° cation to a 3° (as shown in the example), it often happens quickly. It will also happen if a 1° becomes a 2° or 3°. So here are some ways to test mechanism 1 – kinetics, stereochemistry, rearrangement.

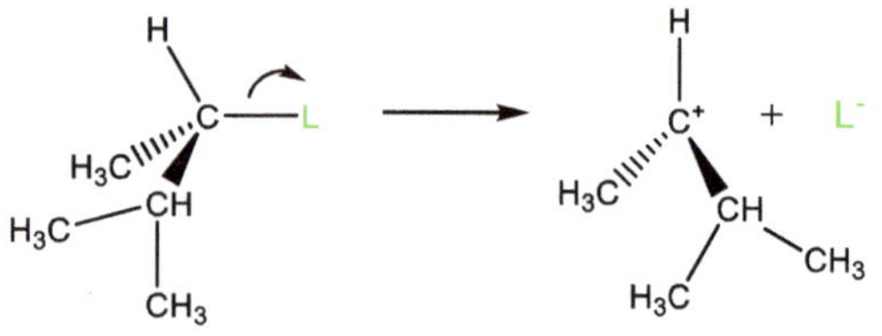

Kinetics:
1st order - Rate = k[R-L]
Rate Determining Step is
Loss of Leaving Group

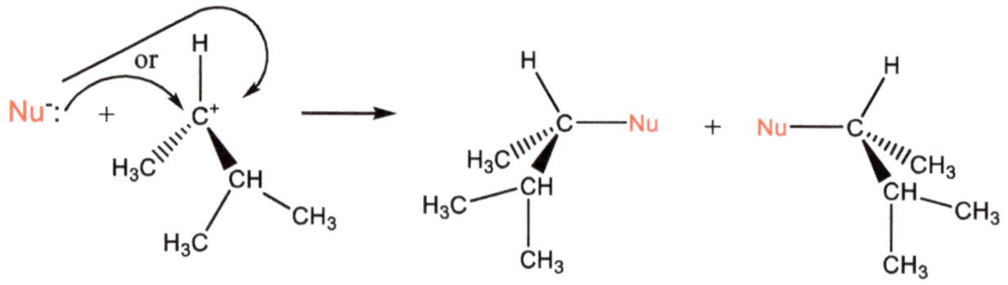

Stereochemistry:
Racemization -
Carbocation is
Achiral

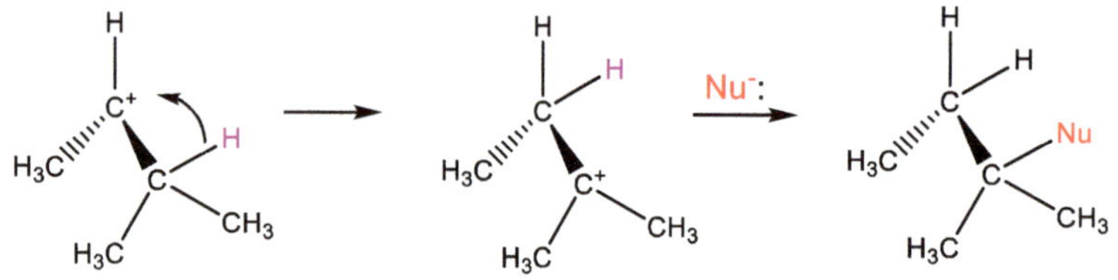

Rearrangement:
2° Cations can
shift -H or -R
to make 3° Cations

In mechanism (2), there is only one step so the rate depends on both the substrate and the nucleophile concentrations: Rate = k[R-L][Nu]. This is a BIMOLECULAR or SECOND ORDER rate. There are no intermediates. There is a TRANSITION STATE – the arrangement of atoms at the maximum energy during the single step. This arrangement only exists for a very, very short time (~10^{-15} sec!) so it cannot be studied directly, and the expected structure is usually drawn in brackets to represent this transitory nature. The double dagger (‡) is the symbol for the transition state. The Nucleophile is required to introduce its lone pair on the opposite side from the leaving group – donating the lone pair into what starts as the antibonding orbital but turns into the bonding orbital as the bond to the leaving group breaks. This requires that if the carbon where reaction occurs is a stereogenic center it must react with INVERSION. Again – rate and stereochemistry are tests. Since there is no carbocation intermediate, rearrangements are unlikely, so the absence of rearrangements is significant.

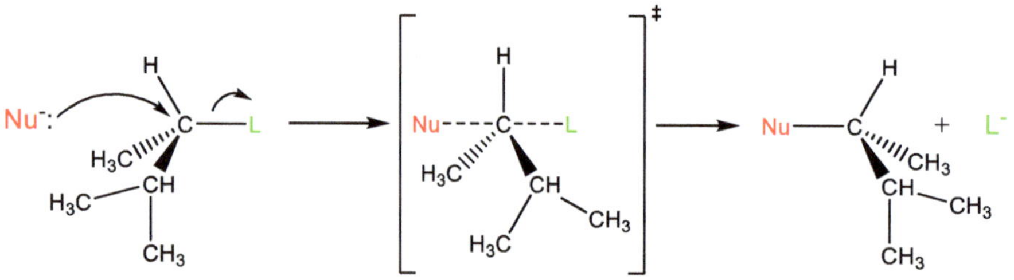

Transition State "‡"

Kinetics:
2nd Order
Rate = k[R-L][Nu]

Stereochemistry:
Inversion - Nu
must come from
side opposite to L

66

Mechanism 3 would also be second order, but it is not clear what the predicted stereochemistry would be. Mechanism 3 requires the formation of a pentavalent carbon anion intermediate, a structure for which there is no evidence. Various attempts to test for the formation of this pentavalent (5 bonds to carbon) intermediate have always failed, so Mechanism 3 is considered disproven.

Pentavalent (5 bonds to C) carbon structures are almost never observed. There is no evidence supporting this process

Evidence for both mechanism 1 and mechanism 2 is favorable. Mechanism 1 most often occurs for substrates where the leaving group is attached to a 3° center, occasionally for a 2° center, only very rarely for 1° or methyl centers. Mechanism 2 occurs for substrates where the leaving group is attached to methyl, 1° or 2° centers, never for 3° centers. So the determining factor is primarily the structure of the substrate. The two mechanisms are commonly known as Substitution, nucleophilic, 1st order or S_N1 and Substitution, nucleophilic, 2nd order or S_N2.

3. The most important thing in S_N1 reactions is the CARBOCATION. This mechanism is only followed if cation formation is favorable. That requires the cation to be as low energy (or most stable) as possible. So cation energies/stabilities have been studied. For alkane type structures, 3° cations are lowest energy/most stable, then 2°, then 1°, then methyl. This is explained by recognizing that each alkyl substituent contains a pool of electrons that can be shifted through the bonds to stabilize the positive charge (called an INDUCTIVE effect). So it is best to have 3 alkyl groups, then 2; 1 is not very good, and methyl cations have no alkyl groups at all. To speed the reaction up, a good leaving group helps – that means it is the conjugate base of a strong acid. The nucleophile is not involved and doesn't change the rate – but if there is no nucleophile at all the reaction can't continue. Polar solvents are preferred – methanol and ethanol are good because they usually dissolve the substrate but are very polar. Water is a poor solvent for the substrate but is even more polar. Relatively high temperature are needed to break the bond and form the cation, so around 100 °C is needed for 3° substrates, even higher for 2°. If the product formed is chiral, racemization occurs. The carbocation intermediates are able to REARRANGE – so 1° can become 2° or 3°, 2° can become 3°.

4. The most important thing about S_N2 is STERIC HINDRANCE: how bulky the groups around the reacting carbon are. For S_N2 to occur, the nucleophile has to come close to the reacting carbon. Bulky groups make this approach more difficult. So methyl is fastest (all groups are the smallest possible – H), 1° with one group is usually next, then 2° with two groups. 3° with three groups is too hindered and usually can't react by S_N2. There are special cases where a 1° substrate happens to have one alkyl group that is very large and bulky and interferes with the nucleophile, so reactivity is reduced. A good leaving group also speeds up the S_N2 reaction, but here the nucleophile IS important. High concentrations of nucleophile make the reaction go faster, and good nucleophiles go faster than poor ones. The best solvents are "polar aprotic". Being polar, they allow for dissolving anionic nucleophiles that are better Nu. But protic solvents like water or alcohols will use their polar, positively charged H atoms to hydrogen bond with the nucleophiles' lone pairs of electrons, reducing the reactivity of the nucleophile. So aprotic solvents are better – they can't hydrogen bond and block the nucleophile. The most common polar aprotic solvents are dimethyl sulfoxide (DMSO) and dimethyl formamide (DMF). Modest temperatures are preferred (30-50 °C), since the mechanism has bonds forming and breaking at the same time it doesn't require as

much energy as S_N1. If the product formed is chiral (only possible with 2°), inversion of the original stereochemistry is observed.

	substrate	Leaving group	nucleophile	Solvent	temperature	stereochemistry	rearrangement
S_N1	3o>2o>1o>Me	important	not important	Polar	high	racemization	Yes
S_N2	Me>1o>2o>3o	important	important	Polar aprotic	moderate	inversion	No

5. Nucleophilic substitution at an alkyl center (S_N1 and S_N2) is an important reaction to learn. You should be able to predict whether a reaction goes by S_N1 or S_N2. Always look at the substrate first: 3° always S_N1, methyl and 1° always S_N2, only 2° depends on conditions – solvent, temperature, etc. One main consequence of S_N1 vs. S_N2 for chiral 2° substrates is the stereochemistry – S_N1 gives racemic products, S_N2 inverted (but still stereochemically pure) products. For this reason, S_N2 is preferred for chiral 2° substrates – we maintain the pure chirality, rather than losing it to racemization. What kind of compounds are substrates for S_N1 and S_N2? Haloalkanes – iodides are most reactive, than bromides, than chlorides. Fluoroalkanes aren't very reactive but they will react slowly. Alcohols are not very reactive when neutral, but can react in acid: the –OH is protonated to $-OH_2^+$, which is a better leaving group. Alcohols can be made reactive by forming various esters: carboxylic acid esters are poor substrates, but sulfonic acid esters (the sulfonates we saw in Williamson Ether synthesis) are good substrates (similar in reactivity to halides). We can use water or hydroxide as a nucleophile and make alcohols, alkoxides as a nucleophile and make ethers, ammonia or amines as nucleophiles to make amines (1° amines add a second alkyl group to become 2° amines, 2° become 3°). Cyanide ion (CN^-) is a good nucleophile to make nitriles. Other good nucleophiles are thiols (R-SH) and thiolate anions (R-S⁻), thiocyanate (SCN^-) and azide (N_3^- or NNN^-).

Quick Review

1. Nucleophilic substitution is a reaction where an electron rich atom or group of atoms (a nucleophile) replaces another group of atoms (a leaving group). Leaving groups are "better" when they are the conjugate bases of stronger acids. Nucleophiles are better when larger, when less electronegative, when negatively charged (vs. neutral), and when stronger bases (when comparing nucleophiles with the same nucleophilic atom).

2. The mechanism of a reaction is the detailed series of steps from reactants to products. Mechanisms can never be fully proven. There are two mechanisms that sometimes occur for nucleophilic substitution – S_N1 and S_N2.

3. The S_N1 mechanism occurs when the substrate breaks a bond into a carbocation and the leaving group, and in a subsequent step the carbocation combines with the nucleophile. The substrate is most reactive when 3°, and the planar carbocation intermediate gives racemization of chiral products.

4. The S_N2 mechanism occurs when the nucleophile makes a bond to the substrate at the same time as the bond to the leaving group breaks – all in a single step. The substrate is most reactive when methyl, as that is least hindered. The formation of the new bond to the nucleophile is opposite to where the bond to the leaving group breaks, giving inversion of stereochemistry.

5. Nucleophilic substitution is widely used to make many different types of compounds: alcohols, amines, ethers, nitriles, and many others.

14.   Elimination

1. Elimination is another common reaction that occurs for most of the same reactants / substrates that react by nucleophilic substitution. In elimination, the leaving group leaves (just like in substitution) but the substrate also loses H^+ to a base so the product is an alkene. Officially, elimination is any reaction where an organic substrate loses two (or more) atoms and gets a new double bond – but in our examples it is almost always a leaving group and H^+ taken by a base.

Just like with nucleophilic substitution, there are two common mechanisms for elimination. In E1, the first step is that the leaving group breaks away to give the carbocation and a leaving group anion: THIS IS THE SAME FIRST STEP AS IN S_N1! The second step has the base come and take a proton (H^+) to make the double bond. Since the first step is the slow step, the rate depends only on the substrate concentration – first order, unimolecular, just like S_N1: Rate = k[Substrate] Since a carbocation is involved, the usual issues with carbocations will exist: stability, rearrangement, so on.

Kinetics - 1st Order
Rate = k [R-L]
Rate Determining Step is
loss of leaving group

Base removes a proton
From C adjacent to cation
Rearrangement is possible

In E2, there is only one step (just like S_N2). The base comes in and takes a proton (H+) at the same time the leaving group leaves. So the rate depends on both concentrations of Substrate and Base – 2nd Order, bimolecular: Rate = k[Substrate][Base]. No rearrangements occur, but the H and leaving group must be ANTI to each other (the technical term is antiperiplanar).

Kinetics - 2nd Order
Rate = k[R-L][B]
No Rearrangement
H & L must be ANTI

2. In Nucleophilic Substitution, apart from the cation rearrangements, the nucleophile substitutes for the leaving group where it stands. But elimination is a little more complicated. We can take a proton from any neighboring carbon to make a double bond. Depending on the structure, there may be 1, 2, or 3 neighboring carbons to the carbon with the leaving group, and they may all have bonds to H's that could potentially be lost with the leaving group to generate the alkene. This means we often make more than one product. Can we predict the distribution of these products – which will form in the greatest quantity? It turns out the answer is YES, and we call this REGIOCHEMISTRY, the formation of different constitutional isomers where the double bond (or other functional group) is in a different REGION of the molecule. Way back in the 1800's a Russian chemist named Zaitsev (sometimes spelled differently) figured out how to predict the product that forms in greatest quantity – called the

MAJOR PRODUCT. Therefore the predicting rule is called ZAITSEV's RULE. The way he said it: "the major product forms from loss of H$^+$ from the carbon with the fewest H's". There is a new, modern way of saying it: "The major product is the most stable alkene." How do we know the most stable alkene? The more C-C bonds surrounding the C=C double bond, the more stable. So alkenes with four alkyl groups around the C=C are most stable, then alkenes with three alkyl groups, then alkenes with two alkyl groups, and finally the alkenes with only one alkyl group. There are three kinds of alkenes with two alkyl groups: *cis* and *trans* have one alkyl group on each C of the C=C, while "1,1" alkenes have both alkyl groups on the same C. Among these three kinds, 1,1 is most stable, then *trans*, then *cis*. Most eliminations follow the Zaitsev Rule, and favor more substituted alkenes. There are some special reaction conditions that favor less substituted alkenes, these are called "Anti-Zaitsev" or sometimes "Hofmann" from the original developer of a set of conditions to achieve this result.

In the reaction below, the green base makes the most substituted and stable alkene as the major product, while the red and blue bases make less substituted alkenes as minor products.

The Zaitsev Rule: the most substituted alkene is favored and is the major product

When given an elimination reaction as a problem, always remember that there is likely to be a major product and some minor products. The professor will usually expect you to give all products and indicate major vs. minor.

3. The E1 reaction, since it goes through a carbocation, occurs most easily (faster or at lower temperature) for 3° substrates. The reaction will be faster with a better leaving group. The concentration of the base is not critical, and it does not require a very strong base: in many cases, water is a sufficiently strong base. Since it proceeds through ionic intermediates, polar solvents make the reaction faster. High temperatures are necessary to form the ionic intermediate, and higher temperatures favor elimination over substitution. There is no particular stereochemistry for E1: the carbocation intermediates formed are achiral. E1 usually follows Zaitsev's Rule for regiochemistry. The carbocation intermediates will undergo the rearrangements common to carbocations.

While E1 eliminations can occur for 2° and 3° haloalkanes and sulfonate esters, another common elimination procedure that proceeds through E1 starts with alcohols. Here the alcohol is treated with acid and heated, and protonation of the alcohol –OH to –OH$_2^+$ makes it a better leaving group. In acidic solution there cannot be any strong bases for E2, so E1 is the only option. The strength of acid and temperature required vary with substrate structure: 1° alcohols typically are treated with concentrated sulfuric acid and ~180 °C, 2° alcohols with phosphoric acid and 120-140 °C, and 3° alcohols with 10% sulfuric acid and 100-120 °C. Stronger acid and higher temperatures are needed to generate the less stable 1° cation. This is a common reaction to be given in problems, often with rearrangements.

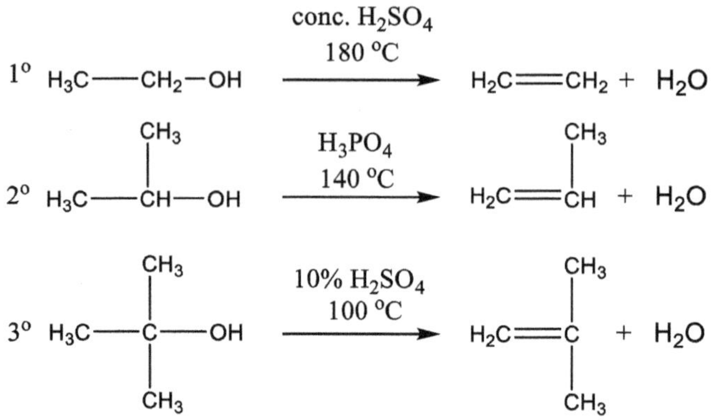

$$1° \quad H_3C\!-\!CH_2\!-\!OH \quad \xrightarrow[\text{180 °C}]{\text{conc. } H_2SO_4} \quad H_2C\!=\!CH_2 + H_2O$$

$$2° \quad H_3C\!-\!\overset{\overset{\displaystyle CH_3}{|}}{C}H\!-\!OH \quad \xrightarrow[\text{140 °C}]{H_3PO_4} \quad H_2C\!=\!\overset{\overset{\displaystyle CH_3}{|}}{C}H + H_2O$$

$$3° \quad H_3C\!-\!\overset{\overset{\displaystyle CH_3}{|}}{\underset{\underset{\displaystyle CH_3}{|}}{C}}\!-\!OH \quad \xrightarrow[\text{100 °C}]{10\% \ H_2SO_4} \quad H_2C\!=\!\overset{\overset{\displaystyle CH_3}{|}}{\underset{\underset{\displaystyle CH_3}{|}}{C}} + H_2O$$

Conditions for acid catalyzed E1 of alcohols

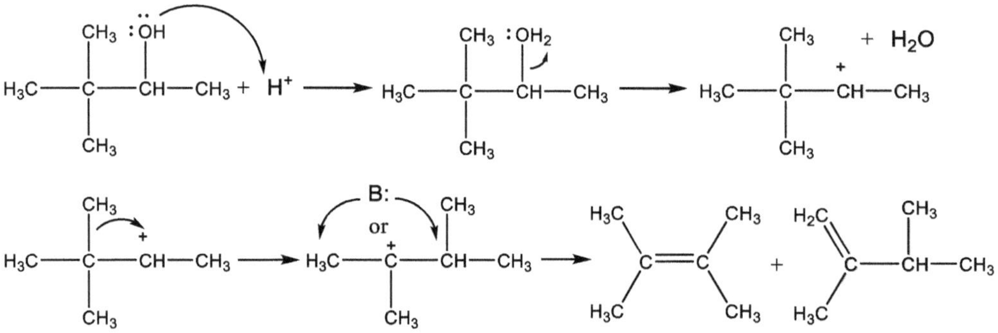

Mechanism: Acid catalyzed dehydration with rearrangement

4. Unlike S_N2, E2 is not usually controlled by steric hindrance. The base doesn't have to access the carbon with the leaving group attached, but only a proton on an adjacent carbon, and these are usually quite accessible. So the weaker bond to the leaving group for 3° substrates makes for faster reactions. Better leaving groups make for faster reactions, and high concentrations of strong base also make the reaction faster. The most common condition is to use sodium ethoxide in refluxing ethanol (~80 °C). Sodium ethoxide is the sodium salt of the conjugate base of ethanol ($CH_3CH_2O^-Na^+$) and is prepared by reacting ethanol with sodium metal. Bases maintain their strength in polar solvents like ethanol, so these are favored – the polar aprotics of S_N2 aren't necessary. Higher temperatures are good, but not too high or E1 might start to occur. E2 usually follows the Zaitsev rule for regiochemistry. There are no rearrangements in E2.

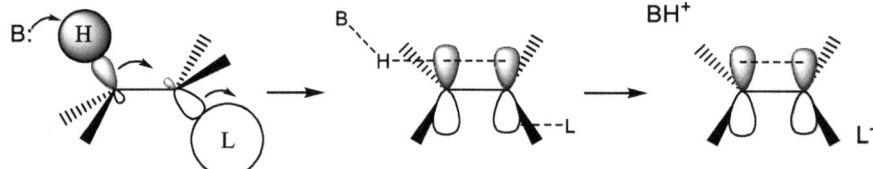

Antiperiplanar elimination in E2 is based on orbital requirements

E2 has a particular stereochemistry that is required as is shown above. This is referred to as "antiperiplanar". It helps to understand this by referring back to the discussion of the conformational analysis of butane. Remember that the "anti" conformation had the two methyl groups 180° apart in the Newman conformation. For E2 to occur, the leaving group and proton (H^+) that are being eliminated have to be in this "anti" arrangement – 180° apart. In that arrangement, the leaving group breaks away while the proton is removed, and the two sp^3 orbitals that are involved in the bonding to the leaving group and proton can easily rotate the 20-30 degrees to become the p orbitals that make the new pi bond. In most E2 reactions, this requirement for the antiperiplanar arrangement isn't obvious. But for certain substrates, the antiperiplanar requirement creates unexpected consequences. One common situation is found in cyclohexanes, where the leaving group and proton must both be axial in order to be antiperiplanar. For substituents on adjacent carbons to both be axial, they must be in a *trans* relationship – if they are *cis*, one will always be equatorial. This can limit the E2 reactions that can occur for cyclohexanes. The other situation where the antiperiplanar arrangement is observed is when the leaving group and proton are both on stereogenic centers. Of the four possible stereoisomers (R,R; R,S; S,R; S,S), two will only give the E product and the other two will only give the Z product.

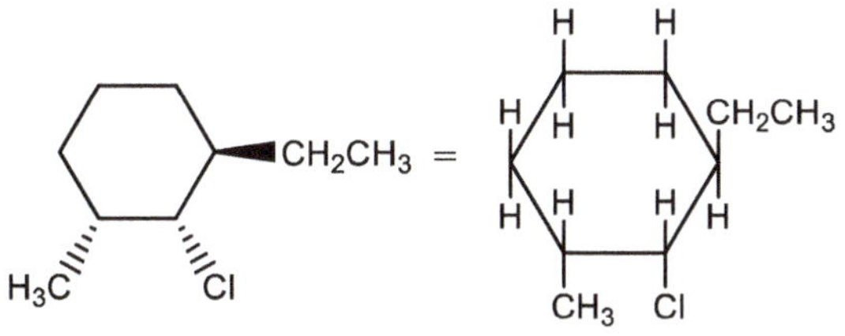

Two ways of drawing a chlorocyclohexane derivative

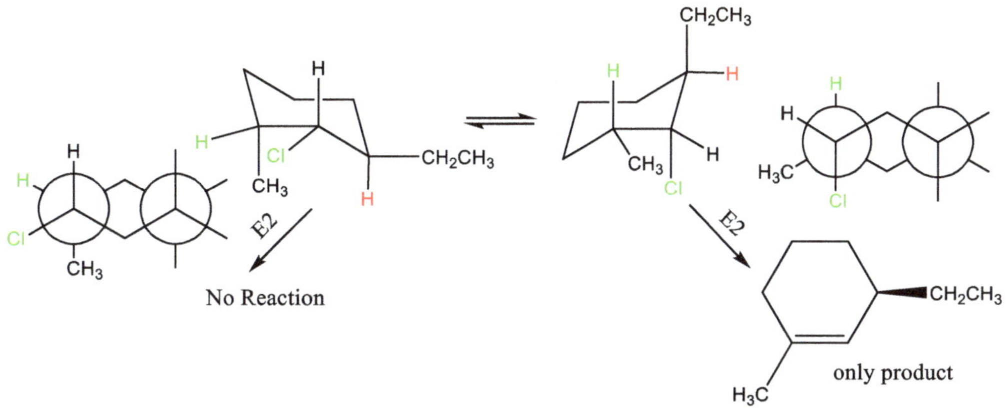

Left chair: When Cl is equatorial, it is anti to the ring C-C bond: no elimination

Right chair: Only when Cl and adjacent H are both axial can elimination occur

Expect to see a problem that tests your knowledge of the antiperiplanar stereochemistry.

	Substrate	Leaving Group	Base	Solvent	Temperature	Stereo-chemistry	Rearrangement	Regio-chemistry
E1	3°>2°>1°	important	Not important	polar	High	none	Yes	Zaitsev
E2	3°>2°>1°	important	important	polar	Medium high	Anti-periplanar	No	Zaitsev

5. The primary function of elimination is the synthesis or preparation of alkenes. Alcohols are commonly available, so the E1 method using alcohols and acid is convenient. The high temperatures and acidic conditions are not always tolerated, and E1 will sometimes lead to rearrangement. E2 conditions give more control, since there is no rearrangement and the antiperiplanar stereochemistry often allows for more control of product distribution, including alkene stereochemistry.

6. Haloalkanes and alkyl sulfonates are substrates for both nucleophilic substitution and elimination. We can often control or predict which will occur based on the substrate and reaction conditions. The best way to ensure nucleophilic substitution is to use S_N2 chemistry. This favors methyl, 1° or 2° substrates, and using low temperatures (25-40 °C) and high concentrations of a strong nucleophile in an aprotic polar solvent. For elimination, E2 is often the choice. This favors 3° substrates, but will work with 2° or 1°. High concentrations of a strong base in a polar solvent at higher temperatures favor E2 – typically sodium ethoxide in refluxing ethanol. Nucleophilic substitution on a 3° substrate is a problem. Only S_N1 is available, and the conditions required (higher temperatures, polar solvents) ALWAYS result in some E1 elimination. E1 is rarely the elimination of choice for haloalkanes and alkyl sulfonates, but is a common method for alcohols in acidic solution. Rearrangement of the cation is always a possibility.

Quick Review

1. ELIMINATION: Elimination is a reaction where groups are lost from adjacent carbons to give a new double bond. In sophomore organic chemistry, the groups are usually H^+ and Cl^- or Br^- as a leaving group with base removing the H^+, or H^+ and OH^- with acid (H^+) catalysis converting OH^- to the better leaving group H_2O.

2. ZAITSEV: Elimination often has regiochemistry – while there is usually only one leaving group, there are multiple H+ on adjacent carbons that can be lost to give different alkenes. The Zaitsev Rule predicts which alkene will be the major product (more of that product than the others). Our modern formulation says the major product is the most stable alkene – which is the most substituted alkene.

3. E1: E1 favors 3° substrates, good leaving groups, high temperatures, and polar solvents. The base is not important. E1 follows the Zaitsev rule for regiochemistry, has no stereochemical requirements, and often displays carbocation rearrangements. Acid catalyzed elimination of alcohols is a common E1 reaction.

4. E2: E2 favors 3° substrates, good leaving groups, high concentrations of strong bases, medium high temperatures, and polar solvents. (Usually sodium ethoxide in refluxing ethanol at 80 °C.) E2 follows the Zaitsev rule for regiochemistry and requires an antiperiplanar arrangement for stereochemistry.

5. ALKENES: Elimination is used to make alkenes.

6. E1/E2 vs S_N1/S_N2: The bimolecular reactions are usually the best choice – S_N2 and E2. S_N2 is favored by methyl, 1° and 2° substrates, low temperatures, high nucleophile concentration, and polar aprotic solvents. E2 is favored by 3° substrates, medium high temperatures, high base concentrations, and polar solvents.

15. Addition

1. Addition is a reaction that is the opposite of elimination. A reagent "X-Y" reacts with an alkene or alkyne to form a product where one carbon of the alkene/alkyne makes a bond to X and the other alkene carbon makes a bond to Y. In many cases, these X-Y reagents are a proton and a nucleophile (H^+ - Nu^-), or more generally an electrophile and a nucleophile (E^+ - Nu^-). Most of the addition reactions we will describe here start with the electrophile, so they are considered "electrophilic addition". An electrophile is the opposite of a nucleophile – where nucleophiles are negative, have lone pairs of electrons, and seek out positive charge, electrophiles are positive, have empty orbitals, and seek out negative charge. In this case, the electrophile seeks out the electrons of the pi bond. Electrophile and nucleophile are equivalent to Lewis acids and bases, except that electrophile and nucleophile are evaluated on kinetics and rates rather than equilibria. Here we will look at addition of HCl and HBr, H_2O and equivalents, and Br_2 as electrophilic additions, and also addition of H_2 and some oxygen reagents which are not electrophilic additions. Addition of unsymmetrical reagents (X-Y) can have REGIOCHEMISTRY – which isomer is formed, and all additions can have STEREOCHEMISTRY: do the adding reagents come from the same side of the double bond (SYN) or opposite sides (ANTI). However, it is possible for additions to have neither regiochemistry nor stereochemistry. We reviewed many of these additions as reactions of alkenes and alkynes and as methods for preparation of alcohols and haloalkanes.

Typical electrophilic addition. The electrophile (usually H^+) adds first,
the nucleophile adds afterwards. The stereochemistry is variable.

2. Addition of HX: The addition of HCl or HBr (generalized as HX) is one of the simplest additions. In the 1800's, a Russian chemist named Markovnikov studied additions (like Zaitsev had studied elimination). He found a certain pattern to the addition of HX, which he stated like this: In the addition of HX, the H adds to the carbon with more H's, and the X adds to the carbon with less H's. This is called MARKOVNIKOV'S RULE. We now have a mechanistic understanding of addition, so we state the Rule a little differently. In the mechanism of addition, the first step is the addition of H^+ to the alkene to make a carbocation. In a second step, the nucleophile (Cl^- or Br^-) adds to the carbocation to complete the addition. The first step determines the REGIOCHEMISTRY of the product – Markovnikov's Rule. So the modern statement of the Rule is that addition of HX proceeds through the more stable carbocation. In most cases, the original statement of the rule and the modern statement are identical. Because the carbocation that is formed is planar, the second step (nucleophilic attack of halide ion) can come from either side: so there is no set STEREOCHEMISTRY of addition for HX. In Markovnikov additions, it is useful to use propene as the alkene substrate, as shown here:

Markovnikov's Rule: H adds to the carbon with more H's (CH_2); Cl adds to the carbon with less H's (CH)

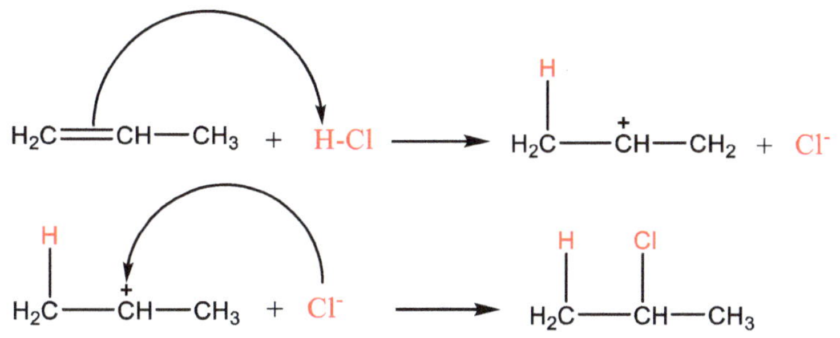

$$H_2C{=}CH{-}CH_3 \; + \; H\text{-}Cl \longrightarrow H_2C{-}\overset{+}{C}H{-}CH_2 \; + \; Cl^-$$

$$H_2C{-}\overset{+}{C}H{-}CH_3 \; + \; Cl^- \longrightarrow H_2C{-}CH{-}CH_3$$

Mechanism explaining Markovnikov's Rule: the first step (addition of H^+) makes the more stable carbocation (2° over 1°), providing a particular regiochemistry (2-chloropropane not 1-chloropropane).

Since addition goes through the carbocation, the typical properties of carbocations hold: 3° are favored over 2° and 1°, and 2° are favored over 1°. If there is a route for rearrangement to a more stable carbocation, the rearrangement usually occurs rapidly and the rearranged product is found to be the major product or the only product. Markovnikov additions – with or without rearrangement – are common exam questions.

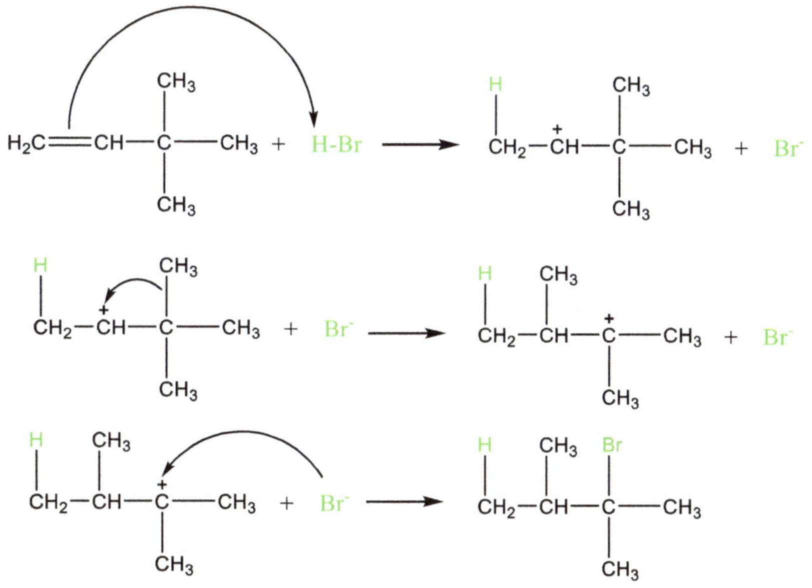

$$H_2C{=}CH{-}\underset{CH_3}{\overset{CH_3}{C}}{-}CH_3 \; + \; H\text{-}Br \longrightarrow CH_2{-}\overset{+}{C}H{-}\underset{CH_3}{\overset{CH_3}{C}}{-}CH_3 \; + \; Br^-$$

$$\overset{+}{C}H_2{-}CH{-}\underset{CH_3}{\overset{CH_3}{C}}{-}CH_3 \; + \; Br^- \longrightarrow CH_2{-}CH{-}\underset{CH_3}{\overset{CH_3}{\overset{+}{C}}}{-}CH_3 \; + \; Br^-$$

$$CH_2{-}CH{-}\underset{CH_3}{\overset{CH_3}{\overset{+}{C}}}{-}CH_3 \; + \; Br^- \longrightarrow CH_2{-}CH{-}\underset{CH_3}{\overset{CH_3\;\;Br}{C}}{-}CH_3$$

Markovnikov addition with rearrangement.

Alkynes can also be subjected to addition. With addition of one equivalent of HX to a terminal alkyne, the addition follows Markovnikov's rule to make a 2-haloalkene. Addition of a second equivalent of HX makes the 2,2-dihaloalkene. Markovnikov's rule doesn't apply to most internal alkynes since both cations are identically substituted.

3. Addition of H_2O and equivalents: another classic addition is the addition of water (H-OH for our purposes). Unlike HCl and HBr, water is not sufficiently acidic to add H^+ on its own. An external source of H^+ - an acid catalyst – must be added. The best catalysts are those that have counterions that are not nucleophiles: so HCl and HBr are not so good. The usual choice is dilute H_2SO_4: HSO_4^- is only very weakly nucleophilic, and any sulfates that do form can be hydrolyzed with water to give the alcohol. The technical name for this reaction is ACID CATALYZED

HYDRATION. It follows a similar mechanism to the addition of HX, so it follows Markovnikov's Rule and suffers from rearrangements.

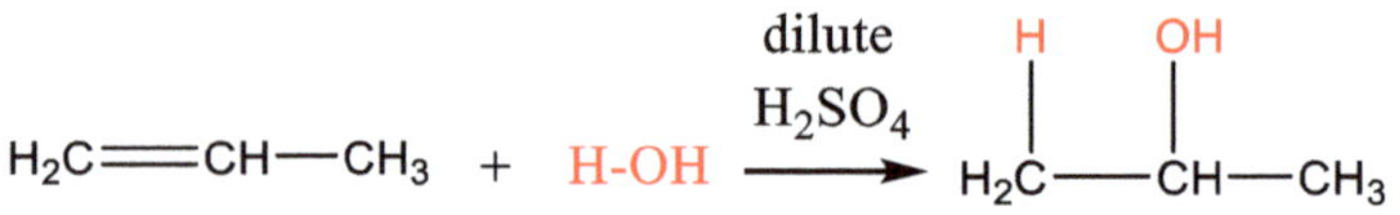

$$H_2C=CH-CH_3 \; + \; H\text{-}OH \xrightarrow[H_2SO_4]{\text{dilute}} \underset{H_2C-CH-CH_3}{\overset{H \quad OH}{|\quad\;\;|}}$$

Markovnikov hydration of propene. The dilute H_2SO_4 is sometimes simply written as "H^+".

Alkynes can also react by hydration. The product is an "enol" – a compound with a hydroxyl group on the alkene double bond. These structures are usually unstable relative to the "keto" isomers – carbonyl compounds with C=O bonds, and rapidly equilibrate to the keto isomers. So hydration of terminal alkynes results in ketones at C2.

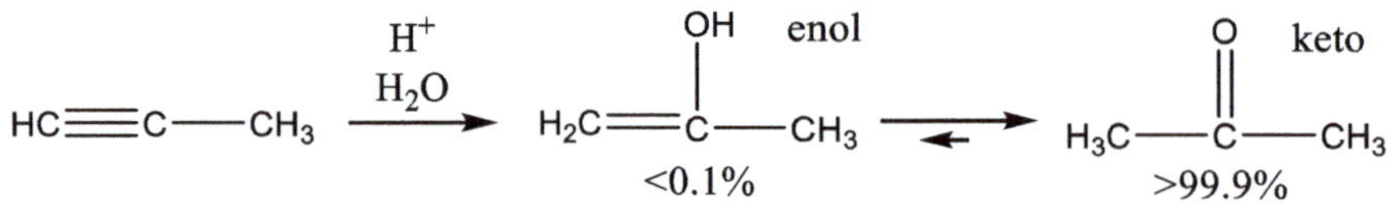

$$HC\equiv C-CH_3 \xrightarrow[H_2O]{H^+} \underset{<0.1\%}{\overset{OH \quad \text{enol}}{H_2C=C-CH_3}} \rightleftharpoons \underset{>99.9\%}{\overset{O \quad \text{keto}}{H_3C-C-CH_3}}$$

Hydration and isomerization of alkynes

Alcohols are the product of hydration. They are useful starting materials in many syntheses. For this reason, chemists sometimes want alcohols that are not the product of Markovnikov addition or are not rearranged to the most stable carbocation. So techniques have been developed to add water INDIRECTLY. After the reaction, it looks like water was added, but water was not the source of both the H and OH. There are two commonly discussed methods: one gives Markovnikov addition but prevents rearrangements, the other gives the opposite regiochemistry, so it is called ANTI-Markovnikov addition.

Markovnikov addition without rearrangements: Oxymercuration-Demercuration. To accomplish effective addition of water with no rearrangement, a different electrophile must be used. The most common one is Hg^{2+}. In the OXYMERCURATION step, the alkene is treated with Mercury (II) Acetate, usually written $Hg(OAc)_2$, along with water and often a co-solvent like THF (since the $Hg(OAc)_2$ won't dissolve in organics but the alkene won't dissolve in water). This results in an organometallic compound, which is the Markovnikov alcohol with Hg(OAc) where the H would have added. Because the Hg^{2+} ion is so large (atomic number 80), it effectively prevents rearrangement (we will see how below). In the DEMERCURATION (or reduction) step, this organometallic is treated with sodium borohydride, which reduces the mercury from the 2+ cation to mercury metal (Hg^o), and replaces it with H. This gives the Markovnikov product as shown.

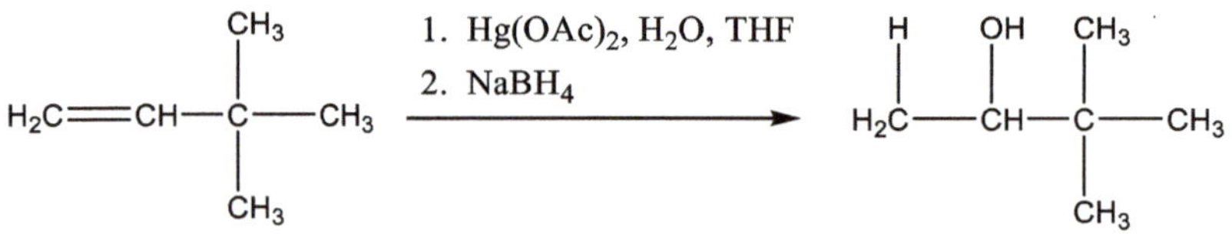

$$H_2C=CH-\underset{\underset{CH_3}{|}}{\overset{\overset{CH_3}{|}}{C}}-CH_3 \xrightarrow[\text{2. } NaBH_4]{\text{1. } Hg(OAc)_2,\, H_2O,\, THF} \underset{H_2C-CH-\underset{\underset{CH_3}{|}}{\overset{\overset{CH_3}{|}}{C}}-CH_3}{\overset{H \quad OH \quad CH_3}{|\quad\;\; |}}$$

Here are the details:

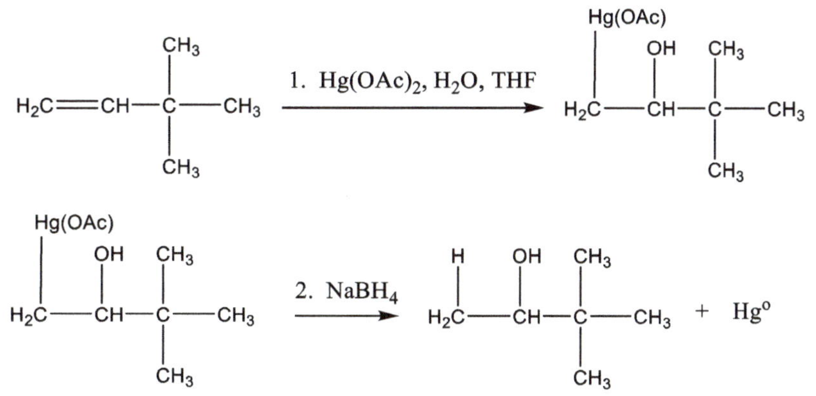

CH_3 reaction scheme:

H_2C=CH—C(—CH_3)(—CH_3)—CH_3 → 1. $Hg(OAc)_2$, H_2O, THF → H_2C—CH—C(CH_3)—CH_3 with OH and Hg(OAc)

$Hg(OAc)$ | OH CH_3 — H_2C—CH—C—CH_3 — 2. $NaBH_4$ → H OH CH_3 H_2C—CH—C—CH_3 + Hg^o

Anti-Markovnikov addition: Hydroboration-Oxidation. Sometimes chemists want to make the other regioisomer: 1-propanol instead of 2-propanol, for example. Regular electrophilic addition won't make this product. The method chemists have developed is to react the alkene with borane (BH_3). Borane is an odd substance (along with being pyrophoric and highly toxic), but the important part here is that boron is LESS electronegative than hydrogen, so the B-H bond is polarized as B^+-H^- : B is the electrophile and H is the nucleophile! So when we compare adding BH_3 to H_2O, the H will add to the *more* substituted carbon for BH_3 and the *less* substituted carbon for H_2O. The product of addition of BH_3 is an alkylborane. Since there are 3 H's on BH_3, it can add to 3 alkenes, but an excess is often used. When the alkylborane is oxidized in base (H_2O_2, OH^-), the boron is replaced with a hydroxyl to make an alcohol. (The boron is oxidized to borate ion BO_3^{-3}) The net result is anti-Markovnikov addition, as shown here.

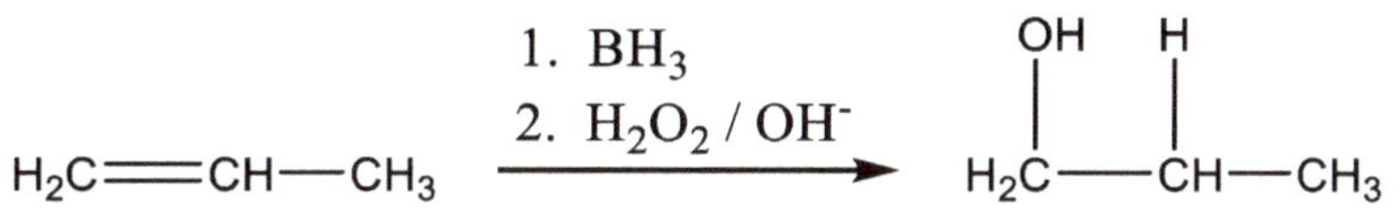

H_2C=CH—CH_3 → 1. BH_3 2. H_2O_2 / OH^- → OH H H_2C—CH—CH_3

There are a few more important details for hydroboration-oxidation. First, addition of BH_3 is a SYN addition: the B and H add from the same side (it is believed that they add in a single step). Second, oxidation of the alkylborane to alcohol proceeds with RETENTION of stereochemistry. This makes the stereochemical outcome of hydroboration-oxidation highly predictable.

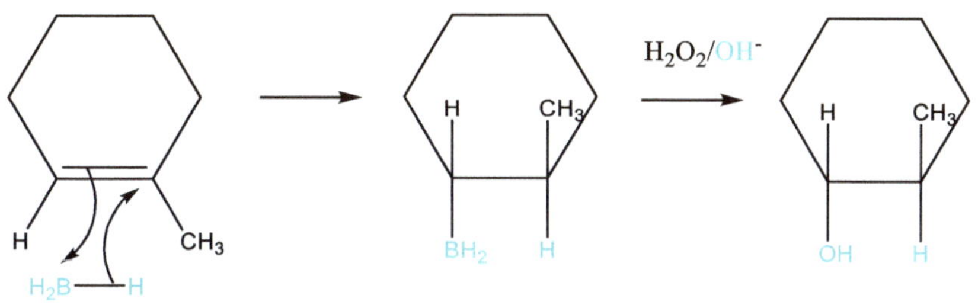

Hydroboration-oxidation gives Anti-Markovnikov addition
Addition is SYN, OH replaces B with Retention of Stereochemistry

Hydroboration-oxidation applied to terminal alkynes gives the enol with the alcohol at C1, which then isomerizes to an aldehyde.

Addition to make alcohols, whether acid catalyzed hydration, oxymercuration-demercuration or hydroboration-oxidation, are common questions on exams. Often the same alkene is subjected to several of these reactions – you should recognize that the products will be isomers.

4. Addition of Halogens: Cl_2 and Br_2. When treated with the halogens Cl_2 and Br_2, alkenes add these elements across the double bond. Because the two added atoms are identical, there can be no regiochemistry – so Markovnikov's Rule doesn't apply.

$$H_2C=CH-CH_3 \ + \ X_2 \longrightarrow XH_2C=CHX-CH_3$$

This addition does have a stereochemistry however – the addition is ANTI: the two halogens add from opposite sides of the double bond. The explanation for ANTI addition lies in the mechanism. In the first step, the alkene double bond reacts with the halogen molecule, forming a bridged halonium ion and displacing the other halogen as halide. Because the Cl and Br atoms are large, they are capable of forming the bridged ion (a similar bridged ion forms in the oxymercuration described above with the large Hg ion). In the second step, the halide ion acts as a nucleophile – because of the bridging, it can only attack from the opposite side, resulting in ANTI addition.

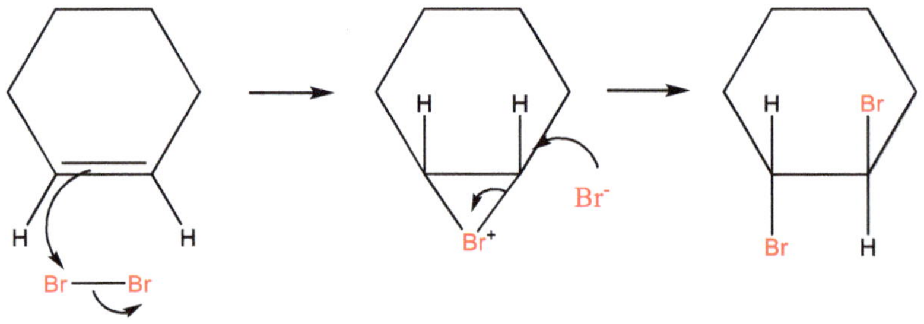

Mechanism of halogenation, explaining ANTI stereochemistry of addition. Bromine is shown, but chlorine acts identically.

Traditionally halogenation was performed with carbon tetrachloride as solvent – but this was banned as an ozone-depleting chemical about 25 years ago. Now it is usually performed in chloroform, or a hydrocarbon like hexane. It is also possible to perform the reaction in water – but then the nucleophile in the second step is water, and the product is called a halohydrin:

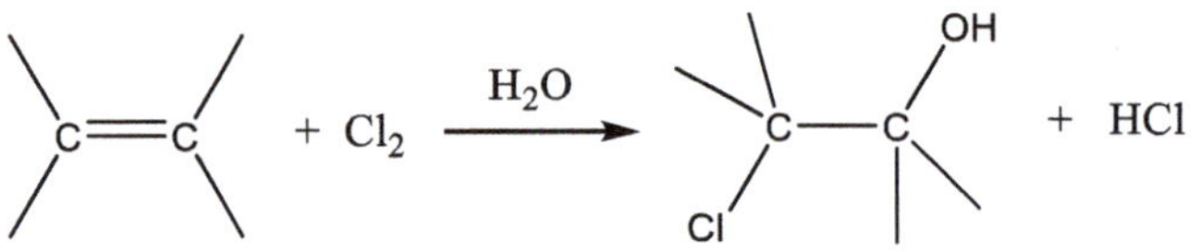

Formation of a halohydrin by halogen addition in water. Chlorine is shown, but bromine acts identically. This follows Markovnikov's Rule: the halogen is the electrophile and the hydroxyl is the nucleophile.

Halogenation of alkynes with 1 equivalent of the halogen gives dihalo alkenes with the halogens trans to each other. Using excess halogen (beyond 2 equivalents) results in two halogen atoms on each carbon of the former triple bond.

Halogenation is a common exam question, particularly in situations where the ANTI character of the addition is relevant to stereochemistry – work out the products for ANTI bromination of cis 2-butene and trans 2-butene, for example. This is a classic stereospecific reaction.

5. Oxidations. Alkenes/alkynes can be oxidized. These are all relatively similar, but the choice of reagent determines the ultimate products. In organic chemistry, we consider reactions to be oxidations when the number

of bonds to oxygen increase. One type of oxidation is dihydroxylation, the addition of a hydroxyl to each carbon of the double bond. There are two methods of dihydroxylation, one that results in SYN dihydroxylation and the other results in ANTI dihydroxylation. SYN dihydroxylation is most often accomplished with the reagent osmium tetroxide (OsO_4). Osmium tetroxide adds the two oxygens simultaneously from the same side, making an intermediate which is a cyclic "osmate ester". The reaction is very efficient, but OsO_4 is highly toxic and expensive.

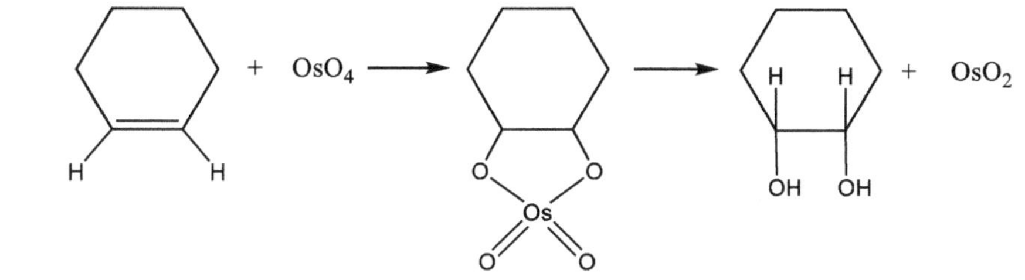

SYN hydroxylation with OsO_4

ANTI dihydroxylation is a two step process: first the alkene is treated with the reagent MCPBA (m-chloroperbenzoic acid) which converts the alkene to an epoxide (3 membered ring ether). Then the ring is opened with water, using either acid or base catalyst, to give the ANTI dihydroxyl product as a racemic mixture. An older way to make the epoxide intermediate is from a halohydrin with base.

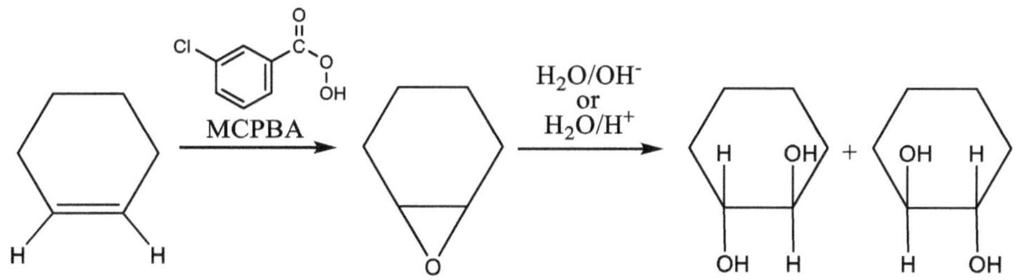

ANTI dihydroxylation through the epoxide.

Dihydroxylation is not used with alkynes.

Oxidation can also cleave the double bond. The most common way to do this is with ozone gas (O_3) followed by treatment with zinc or dimethyl sulfide. This results in a pair of carbon oxygen double bonds C=O where the alkene C=C bond was, with ketone and aldehyde products. This is sometimes done with potassium permanganate in base. The products here are ketones, carboxylic acids instead of aldehydes, or carbon dioxide (for the CH_2 of terminal alkenes). A frequent use of oxidative cleavage is analytical, to locate a double bond: if we have a long chain alkene like $C_{18}H_{36}$, it can be hard to tell which of the many isomeric octadecenes we have. After oxidative cleavage, we can identify the chain length of the two aldehyde or acid products and establish where the double bond was in the starting material.

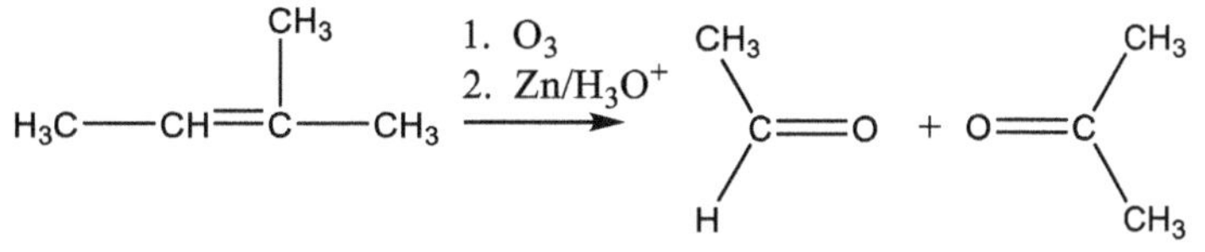

Oxidative cleavage with ozone: aldehyde and ketone products

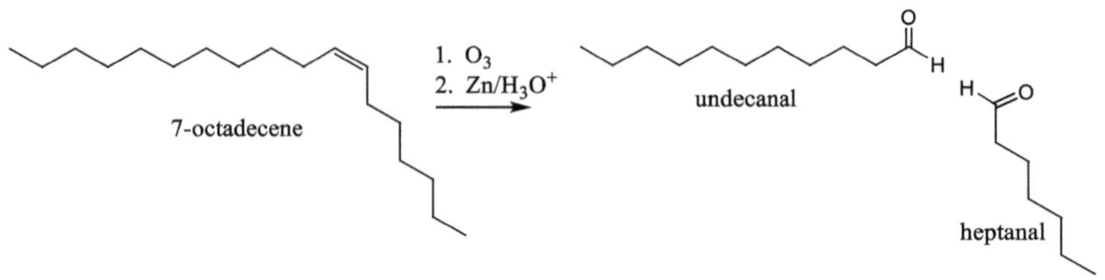

7-octadecene
1. O_3
2. Zn/H_3O^+
undecanal
heptanal

Analysis by cleavage: 7-octadecene gives heptanal & undecanal (C_{11}) products; 6-octadecene would give hexanal and dodecanal (C_{12}), etc.

Oxidation of alkynes gives carboxylic acids - or CO_2 for the CH of terminal alkynes, both with ozone and potassium permanganate.

Questions about oxidation are often asked in the context of the stereochemistry of the SYN and ANTI dihydroxylations, or in the context of the alkene analysis. Be aware that for a cycloalkane, cleavage will produce a molecule with two carbonyl ends (ketone, aldehyde or acid depending on structure and reagent)!

5. Addition of hydrogen (H_2). *Hydrogenation* of alkenes is a very important reaction in terms of industrial applications. In the presence of certain metal catalysts, hydrogen gas and alkenes react to form the alkane: H-H adds across the double bond. This is considered to be a REDUCTION, since the alkene carbons gain H's. Sometimes it is easier to prepare the alkene (or alkyne), then reduce to the alkane, rather than trying to prepare the alkane directly. The metals used as catalysts are most often platinum, palladium, and nickel: Pt is best (but most expensive), Ni least effective (but cheapest).

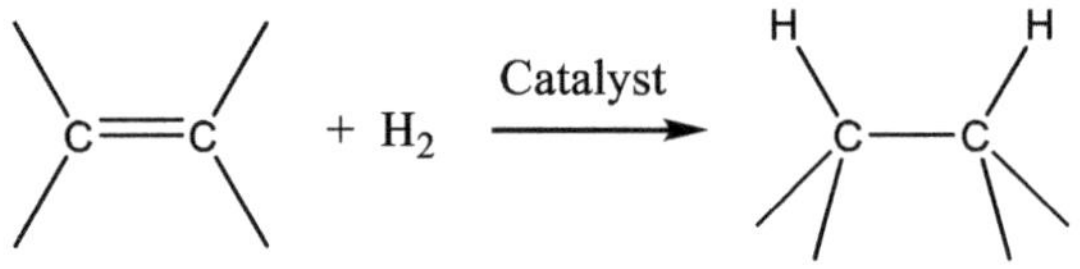

This is a SYN addition – the two H atoms add from the same side. It is understood as occurring on the metal surface: H_2 adsorbs to the surface and breaks into H atoms, and the alkene approaches the surface and picks up both H atoms at once. Since only a thin layer of metal is needed, the expensive Pt and Pd catalysts are often used as thin layers coated onto inexpensive solid supports (silica, graphitic carbon).

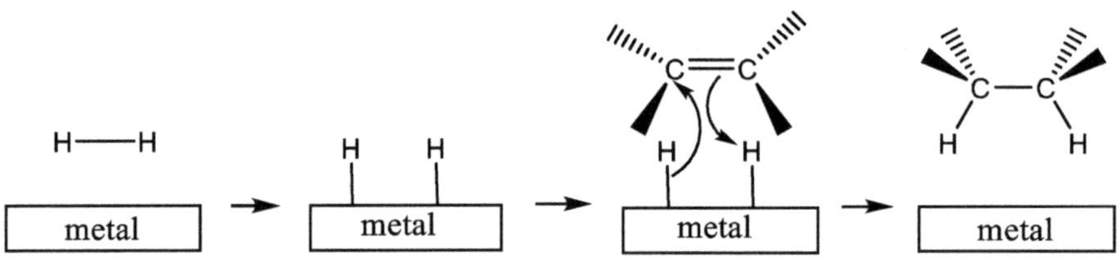

Syn addition of hydrogen from metal surface.

This hydrogenation is often discussed in terms of fats and oils. Fats and oils are tri-esters of glycerol (1,2,3-propanetriol) and fatty acids (long chain carboxylic acids, 10-30 carbons). There are always mixtures of different fatty acids. If the fatty acids are longer chains and alkanes (SATURATED), then the substance is solid at room temperature and called a fat. If enough of the fatty acids have double bonds (UNSATURATED), they are bent and disordered – liquid at room temperature, and called an oil. If you take vegetable oil (unsaturated) and fully hydrogenate it, it becomes "vegetable shortening", a solid at room temperature. Oxygen reacts with the double bonds, oxidatively cleaves them to the acids and makes unsaturated oils taste bad – "rancid". This does not happen to saturated fats – they do not go "bad" nearly as quickly. But the unsaturated fats are considered to be healthier – so there is a tension between shelf life and healthiness. One solution was to only partially hydrogenate the vegetable oil – only hydrogenate some of the double bonds so it doesn't go rancid as quickly. But when a limited supply of hydrogen is used, the hydrogenation can run in reverse – it can remove hydrogens from alkanes to make alkenes. But it makes the more stable TRANS isomers – the naturally occurring unsaturated fats always have CIS double bonds. The resulting TRANS fats are supposedly just as unhealthy (or even less healthy!) than the saturated fats. As mentioned before, if you see "Partially Hydrogenated Vegetable Oil" on the label, the food has trans fats in it: but if it is less than 0.5 g per serving, the rules allow them to round down to "0 g trans fats".

Hydrogenation is also a common reaction for exam questions – frequently involving the SYN stereochemistry.

7. Uses of addition. Addition is very useful as a preparative method. In particular, the different ways to effectively add water (Markovnikov and Anti-Markovnikov) with their attendant stereochemistries provide a very flexible approach to the synthesis of alcohols, and alcohols are then starting materials for many other compounds. Hydrohalogenation and halogenation are also useful in preparing compounds and starting materials. Hydrogenation is also a very frequently used reaction – preparation of alkenes or alkynes is easier than direct preparation of alkanes, so when alkanes are desired, routes through the alkenes or alkynes are common. Oxidation of alkenes was once very common as an analytical method, but the development of modern analytical methods make it less common – it is only rarely used to prepare compounds.

Quick Review

1. ADDITION. Addition reactions occur when a reagent (X-Y) adds across a pi bond (C=C) to make a product (X-C-C-Y). The electrophilic part of the reagent usually adds first – often it is H^+.

2. HYDROHALOGENATION. Addition of HX (HCl or HBr). Follows Markovnikov's Rule: H^+ adds to make the more stable carbocation. This controls regiochemistry. Carbocation rearrangements can occur.

3. HYDRATION. Addition of H-OH (H_2O). Simple acid catalyzed follows Markovnikov and rearrangements occur. Alternate methods allow Markovnikov with no rearrangements (oxymercuration – demercuration) and Anti-Markovnikov (hydroboration-oxidation – has SYN stereochemistry).

4. HALOGENATION. Addition of X_2 (Cl_2 or Br_2). ANTI addition through bridged intermediate.

5. OXIDATION. Dihydroxylation can be SYN (OsO_4) or ANTI (MCPBA-hydrolysis). Oxidative cleavage gives carbonyl derivatives where pi bond was in starting material, often used for analysis.

6. HYDROGENATION. SYN addition of H_2 with metal catalyst. Makes alkanes, partial hydrogenation can result in isomerization of *cis* to *trans* alkenes.

7. USES. The most common uses of addition are in the preparation of other compounds from alkenes and alkynes.

16.   Radicals

1. What are radicals? Earlier we saw the reactive intermediates known as carbocations. These are missing a bond and have an empty orbital with a +1 charge. They are thought of as planar, sp^2 hybridized centers and frequently rearrange to more stable carbocations. Another type of reactive intermediate are known as radicals. These are also missing a bond, but have a singly occupied orbital and are neutral – charge is 0. There is some argument about whether they are planar, sp^2 hybridized centers or tetrahedral, sp^3 hybridized centers that undergo rapid inversion. Either way, they are achiral and usually give racemic products.

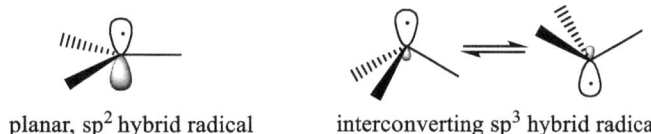

planar, sp^2 hybrid radical   interconverting sp^3 hybrid radical

Radicals form when bonds break homolytically (each fragment gets one electron) rather than heterolytically (one fragment gets both electrons, one fragment gets none). There are certain types of reactions that tend to be associated with radicals: homolytic cleavage, atom abstraction, radical addition to alkenes, recombination.

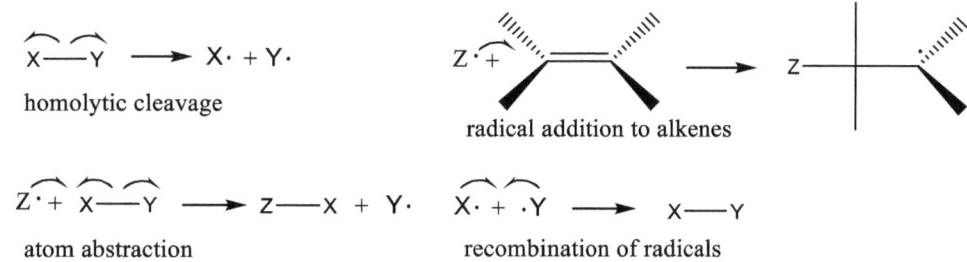

2. Chain reactions. One characteristic of radical reactions is that they are often chain reactions. Since a radical is an odd electron species, when it reacts with an even electron compound (pretty much all organic compounds!), the products have an odd electron – one of them is a radical! (This is simple math: odd number + even number = odd number.) Once radicals have been generated, the only way to be rid of them is when two radicals react together: odd + odd = even. So in a typical radical reaction there is INITIATION – the generation of radicals by homolytic cleavage (the chain begins); PROPAGATION – reactions where radicals are consumed but new radicals are made (the chain continues or propagates); and TERMINATION – reactions between radicals that end (terminate) the chain. One intitiation event might lead to hundreds or thousands of propagation reactions as the chain continues. You should understand the chain reaction mechanisms and the types of steps involved.

3. Radical halogenation. The simplest radical reaction is the radical halogenation of alkanes, which is usually first shown with methane. Chlorine and bromine (X_2) will undergo homolysis with either heat or UV light to give chlorine or bromine atoms, which are odd electron species, therefore essentially radicals X•. This is the INITIATION step. Halogen atoms in turn can abstract hydrogens (H•) from alkanes, generating alkyl radicals (methyl radicals with methane). The alkyl radicals then abstract halogen from another X_2, giving the haloalkane and a new halogen atom that starts a new cycle. These are the PROPAGATION steps. Eventually, either two alkyl radicals or two halogen atoms or one radical and one atom combine to end the chain – this is the TERMINATION step. You should be able to identify these steps.

$$\text{OVERALL REACTION: } CH_4 + Cl_2 \xrightarrow{h\nu} CH_3Cl + HCl$$

$$\text{INITIATION: } Cl\text{—}Cl \xrightarrow{h\nu} Cl\cdot + Cl\cdot$$

$$\text{PROPAGATION: } Cl\cdot + CH_4 \longrightarrow H\text{—}Cl + CH_3\cdot$$

$$CH_3\cdot + Cl\text{—}Cl \longrightarrow H_3C\text{—}Cl + Cl\cdot$$

$$\text{TERMINATION: } CH_3\cdot + CH_3\cdot \longrightarrow H_3C\text{—}CH_3$$

$$Cl\cdot + Cl\cdot \longrightarrow Cl_2$$

$$CH_3\cdot + Cl\cdot \longrightarrow H_3C\text{—}Cl$$

If we use equal amounts of halogen and methane this reaction is not the best. As the reaction proceeds, the amount of chloromethane increases until chloromethane is high enough in concentration to compete with methane in the first propagation step – this generates a chloromethyl radical, and dichloromethane as product. Further on in the reaction, dichloromethane reacts to give trichloromethane. Near the end of the reaction, trichloromethane reacts to give carbon tetrachloride. In the end, the major product is chloromethane but there are minor products of dichloromethane, trichloromethane and carbon tetrachloride. Some unreacted methane is left over. If we want to maximize the chloromethane to near 100%, we can use a large excess of methane – 10-fold or more. Then there is never enough chloromethane to compete effectively with the methane, so only chloromethane is formed as product. Using an excess of an alkane is an effective way to provide only the monohalogenation product – one halogen is added to the alkane. If we want to completely halogenate, then we need to use an excess of the halogen – for methane, at least 4 fold, for ethane at least 6 fold, and so on.

$$1{:}1 \quad CH_4 + Cl_2 \xrightarrow{h\nu} CH_3Cl + CH_2Cl_2 + CHCl_3 + CCl_4 + HCl$$

$$\text{excess } CH_4 \quad CH_4 + Cl_2 \xrightarrow{h\nu} CH_3Cl + HCl$$

4. Chlorine vs. Bromine. We find that radical halogenation with chlorine and bromine give different results. If we run a monohalogenation with chlorine, we get approximately equal amounts of all possible monochlorination products. If we run a monohalogenation with bromine, we get selective bromination: bromination at $3°$ positions is most favored, then bromination at $2°$ positions, and bromination at $1°$ positions is least favored. Why is this? It comes from the energetics of the process. Chlorination is a highly exothermic process, and the abstraction of hydrogen step is exothermic from any type of carbon. Therefore the rate of abstraction is nearly the same for all positions for chlorine and chlorination is nonselective. On the other hand, bromination is either mildly exothermic or mildly endothermic, and the hydrogen abstraction step is endothermic. This makes the rate of abstraction highly dependent on the type of carbon, so that bromination is selective. For 2-methylbutane, while chlorination gives about equal amounts of the four monochlorination products, bromination gives 95-98% of 2-bromo-2-methylbutane (the $3°$ product), 1-4% of 2-bromo-3-methylbutane (the $2°$ product), and less than 1% of each of the two $1°$ products (1-bromo-2-methylbutane and 1-methyl-3-bromobutane). The difference in selectivity of chlorine and bromine is often the basis of questions about radical chemistry.

Cl$_2$

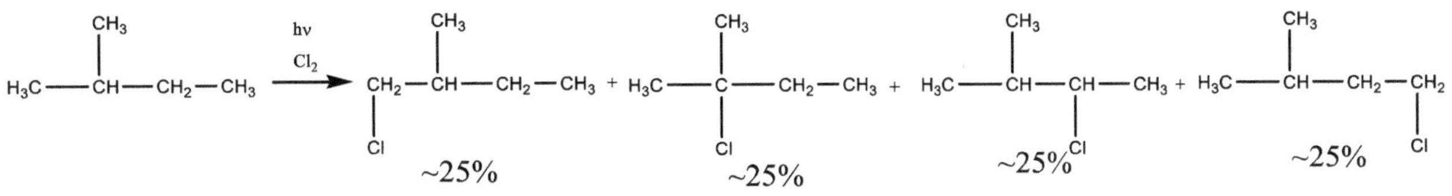

H$_3$C—CH(CH$_3$)—CH$_2$—CH$_3$ $\xrightarrow[\text{Cl}_2]{\text{hv}}$ CH$_2$(Cl)—CH(CH$_3$)—CH$_2$—CH$_3$ + H$_3$C—C(CH$_3$)(Cl)—CH$_2$—CH$_3$ + H$_3$C—CH(CH$_3$)—CH(Cl)—CH$_3$ + H$_3$C—CH(CH$_3$)—CH$_2$—CH$_2$Cl

~25% + ~25% + ~25% + ~25%

Br$_2$

H$_3$C—CH(CH$_3$)—CH$_2$—CH$_3$ $\xrightarrow[\text{Br}_2]{\text{hv}}$ CH$_2$(Br)—CH(CH$_3$)—CH$_2$—CH$_3$ + H$_3$C—C(CH$_3$)(Br)—CH$_2$—CH$_3$ + H$_3$C—CH(CH$_3$)—CH(Br)—CH$_3$ + H$_3$C—CH(CH$_3$)—CH$_2$—CH$_2$Br

<1% + 98% + <2% + <1%

H$_3$C—CH(CH$_3$)—CH$_2$—CH$_3$ + Br· → H$_3$C—C·(CH$_3$)—CH$_2$—CH$_3$ + HBr

3° radical forms fastest - lowest activation energy

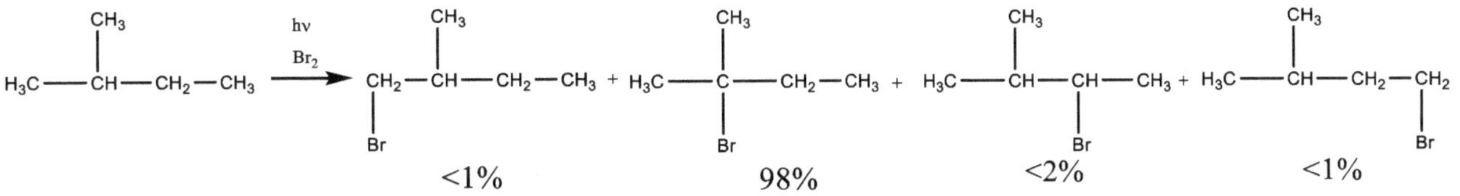

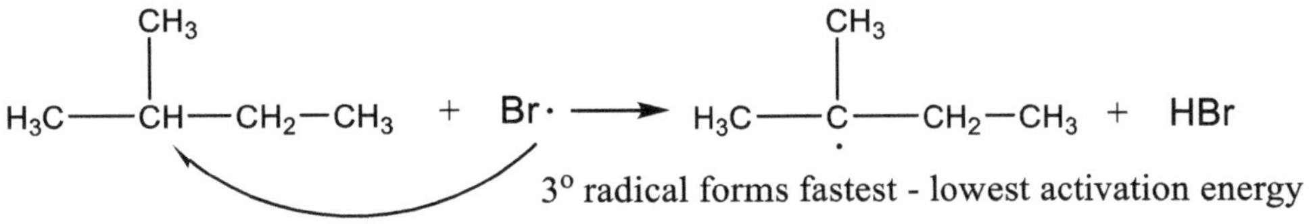

5. Radical Addition and Polymerization. The other major reaction that goes through a radical mechanism is addition. A radical can add to an alkene double bond, which makes a new radical.

Addition of HBr by a radical mechanism is notable because it is Anti-Markovnikov. In the presence of a radical initiator (often a peroxide), the initiator radical abstract H from HBr, giving Br atoms. The Br atoms then add to the alkene so as to make the more stable radical. Since the Br adds first, this gives the Anti-Markovnikov product. This radical then abstracts H from another HBr to release another Br atom, and the chain continues. Combination of Br atoms or carbon radicals terminate the chains. The initiator is most commonly benzoyl peroxide.

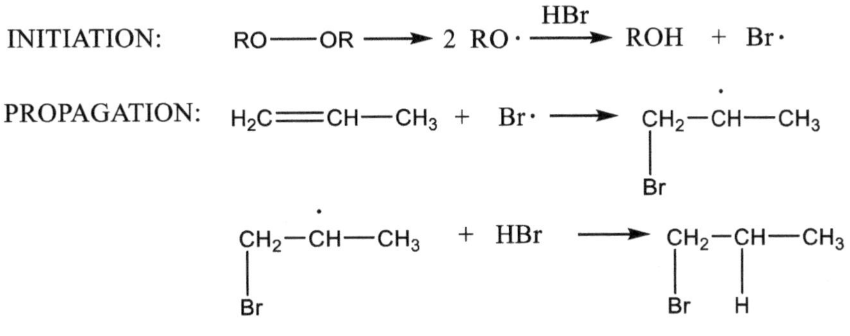

INITIATION: RO—OR → 2 RO· $\xrightarrow{\text{HBr}}$ ROH + Br·

PROPAGATION: H$_2$C=CH—CH$_3$ + Br· → ·CH$_2$—CH—CH$_3$ (with Br)

·CH$_2$—CH—CH$_3$ (with Br) + HBr → CH$_2$—CH—CH$_3$ (with Br and H)

Anti-Markovnikov addition of HBR via radical mechanism (peroxide initiator)

The other common radical addition is polymerization. The alkene is considered a monomer (one unit) that is converted into a polymer (many units). Here the initiating radical (from peroxide or an azo compound) adds directly to the alkene to make an alkyl radical. This alkyl radical adds to another alkene, making a new alkyl radical that is a dimer (two units). This then adds to another to make a trimer and so on. Eventually (after hundreds or thousands of units are added), the radicals either add together (recombination) or one abstracts from the other (disproportionation) to give extremely large molecules – polymers. The process is shown with styrene to make polystyrene, a common polymer used for Styrofoam and other items (plastic cups, etc.). Many objects in our modern world are made by these polymerization processes.

Radical polymerization of styrene (azoisobuyronitrile (AIBN) initiator)

Anti-Markovnikov addition of HBr through a radical process is a more common source of questions than is polymerization at the sophomore level (unless your professor does polymer research – you mean you DON'T KNOW what research your professor does? You should! A lot of times each professor will tend to ask more questions that relate to their own research! As Sun Tzu once said: "Know your enemy".).

Quick Review

1. RADICALS. Radicals are odd electron reactive intermediates.

2. CHAIN REACTIONS. Chain reactions are common in radical chemistry. They are characterized by INITIATION steps that generate radicals, PROPAGATION steps where both starting materials and products contain radicals, these steps are often repetitive, and TERMINATION steps that consume radicals.

3. RADICAL HALOGENATION. Br_2 and Cl_2 will replace the hydrogens of alkanes through a radical mechanism.

4. SELECTIVITY. Addition of Br_2 is selective, favoring 3° over 2° over 1° (the more stable radicals are preferred); addition of Cl_2 is nonselective generating isomeric monochlorinated products nearly equally.

5. RADICAL ADDITION. Radicals add to alkenes. Radical addition of HBr gives the Anti-Markovnikov product. Radical polymerization leads to combination of hundreds or thousands of alkene monomers to form polymers (polystyrene, poly(methyl methacrylate), polyacrylonitrile, polypropylene).

Afterthought

Now we have completed our whirlwind tour of what is usually the first semester of Organic Chemistry, although some of this may have crept into the second semester. This would have taken several hundred pages in a typical organic textbook, but we have completed it in less than 100 – because we have tried to focus on the most important things, to maximize your learning of Organic Chemistry with minimal effort. After you have mastered the material, the Quick Review sections at the end of each topic can be used for a quick refresher course to remind you of the seven or so important points for each topic. Whether you have chosen this to review for a final exam, the MCAT, or some other examination on Organic Chemistry, good luck!

www.ingramcontent.com/pod-product-compliance
Lightning Source LLC
Chambersburg PA
CBHW051021180526
45172CB00002B/435